Internal Combustion Engine: Engineering Fundamentals

Internal Combustion Engine: Engineering Fundamentals

Alison Vaughn

NY RESEARCH PRESS

New York

Published by NY Research Press
118-35 Queens Blvd., Suite 400,
Forest Hills, NY 11375, USA
www.nyresearchpress.com

Internal Combustion Engine: Engineering Fundamentals
Alison Vaughn

International Standard Book Number: 978-1-63238-856-8 (Hardback)

Cataloging-in-Publication Data

Internal combustion engine : engineering fundamentals / Alison Vaughn.
   p. cm.
Includes bibliographical references and index.
ISBN 978-1-63238-856-8
1. Internal combustion engines. 2. Gas producers. 3. Engines. 4. Heat-engines.
5. Engineering. I. Vaughn, Alison.
TJ805 .I58 2022
621.43--dc23

Table of Contents

Preface   VII

Chapter 1   Introduction   1

Chapter 2   Classification of Internal Combustion Engine   7

   a. Reciprocating Engine   7
   b. Diesel Engine   41
   c. Spark-ignition Engine   64
   d. Jet Engine   66
   e. Gas Turbine   78
   f. Rotatory Engine   97

Chapter 3   Components of an Internal Combustion Engine   105

   a. Main Parts of an Internal Combustion Engine 105
   b. Piston   111
   c. Supercharger   116
   d. Turbocharger   128
   e. Air Filter   144
   f. Manifold Absolute Pressure Sensor   146
   g. Starter   148
   h. High Tension Leads   154
   i. Mass Flow Sensor   156

Chapter 4   Fuel and Ignition System   158

   a. Carburetor   158
   b. Fuel Injection   158
   c. Fuel Pump   165
   d. Fuel Filter   171
   e. Ignition System   173
   f. Distributor   177
   g. Wasted Spark   181

Chapter 5	**Exhaust and Cooling System**	**184**
a.	Exhaust System of an Internal Combustion Engine	184
b.	Muffler	189
c.	Catalytic Converter	190
d.	Internal Combustion Engine Cooling	199

Permissions

Index

It is with great pleasure that I present this book. It has been carefully written after numerous discussions with my peers and other practitioners of the field. I would like to take this opportunity to thank my family and friends who have been extremely supporting at every step in my life.

The heat engine where the combustion of a fuel occurs with an oxidizer inside a combustion chamber is known as internal combustion engine. Inside an internal combustion engine, the combustion produces the expansion of the high-temperature and high-pressure gases. This applies direct force to some components of the engine such as turbine blades, pistons, rotor or nozzle. This force moves the components to a distance by transforming chemical energy into mechanical energy. Internal combustion engine can be classified into reciprocating, rotary and continuous combustion. The reciprocating piston engines are the most commonly used engines for land and water vehicles. Rotary engines are used in some aircraft, automobiles and motorcycles. The topics included in this book on internal combustion engine are of utmost significance and bound to provide incredible insights to readers. It outlines the processes and applications of such engines in detail. Those in search of information to further their knowledge will be greatly assisted by this book.

The chapters below are organized to facilitate a comprehensive understanding of the subject:

Chapter – Internal Combustion Engine: An Introduction

The type of heat engine in which the combustion of fuel occurs with an oxidizer in a combustion chamber is known as internal combustion engine. The expansion of high-temperature and high pressure gases produced by combustion provides the propulsive force in this type of engine. This chapter has been carefully written to provide an easy introduction to the varied facets of the internal combustion engine.

Chapter – Classification of Internal Combustion Engine

The internal combustion engine is broadly classified into a number of categories, namely, reciprocating engine, diesel engine, spark-ignition engine, jet engine, gas turbine and rotatory engine. The topics elaborated in this chapter will help in gaining a better perspective about these categories of internal combustion engine.

Chapter – Components of an Internal Combustion Engine

Some of the different components of an internal combustion engine are piston, supercharger, turbocharger, air filter, manifold absolute pressure sensor, starter, high tension leads and mass flow sensor. This chapter closely examines these components of internal combustion engine to provide an extensive understanding of the subject.

Chapter – Fuel and Ignition System

The fuel system is made up of fuel filter, fuel pump, carburetor and fuel injector. The injection system generates a spark to ignite a fuel-air mixture within internal combustion engines. The diverse applications of these components of the fuel and ignition system have been thoroughly discussed in this chapter.

Chapter – Exhaust and Cooling System

The exhaust system in an automobile comprises of the piping which is used to guide reaction exhaust gases away from a controlled combustion inside an engine. The cooling system is employed to keep the temperature of the engine from exceeding the limits imposed by the needs of efficiency and safety. This chapter discusses in detail the diverse aspects of exhaust and cooling systems in an internal combustion engine.

Alison Vaughn

1

Introduction

The type of heat engine in which the combustion of fuel occurs with an oxidizer in a combustion chamber is known as internal combustion engine. The expansion of high-temperature and high pressure gases produced by combustion provides the propulsive force in this type of engine. This chapter has been carefully written to provide an easy introduction to the varied facets of the internal combustion engine.

Internal Combustion Engine

The internal combustion (IC) engine has been the dominant prime mover in our society since its invention in the last quarter of the 19th century. Its purpose is to generate mechanical power from the chemical energy contained in the fuel and released through combustion of the fuel inside the engine. It is this specific point, that fuel is burned inside the work-producing part of the engine, that gives IC engines their name and distinguishes them from other types such as external combustion engines. Although Gas Turbines satisfy the definition of an IC engine, the term has been traditionally associated with spark-ignition (sometimes called Otto, gasoline or petrol engines) and diesel engines (or compression-ignition engines).

Internal combustion engines are used in applications ranging from marine propulsion and power generating sets with capacity exceeding 100 MW to hand-held tools where the power delivered is less than 100 W. This implies that the size and characteristics of today's engines vary widely between large diesels having cylinder bores exceeding 1,000 mm and reciprocating at speeds as low as 100 rpm to small gasoline two-stroke engines with cylinder bores around 20 mm. Within these two extremes lie medium-speed diesel engines, heavy-duty automotive diesels, truck and passenger car engines, aircraft engines, motorcycle engines and small industrial engines. From all these types, the passenger car gasoline and diesel engines have a prominent position since they are, by far, the largest produced engines in the world; as such, their influence on social and economic life is of paramount importance.

The majority of reciprocating internal combustion engines operate on what is known as the *four-stroke cycle*, which is subdivided into four processes: intake, compression, expansion/power and exhaust. Each engine cylinder requires four strokes of its piston which corresponds to two crankshaft revolutions to complete the sequence which lead to the production of power.

The intake stroke is initiated by the downward movement of the piston, which draws into the cylinder fresh fuel/air mixture through the port/valve assembly, and ends when the piston reaches bottom-dead-center (BDC). The mixture is generated either by means of a carburetor (as in conventional engines) or by injection of gasoline at low pressure into the intake port through an eiectronically-controlled pintle-type injector (as in more advanced engines). Effectively, the induction process starts with the opening of the intake valve just before top-dead-center (TDC) and ends when the intake valve (or valves in four-valve per-cylinder engines) closes shortly after BDC. The closing time of the intake valves is a function of the design of the induction manifold, which influences the gas dynamics and volumetric efficiency of the engine, and engine speed.

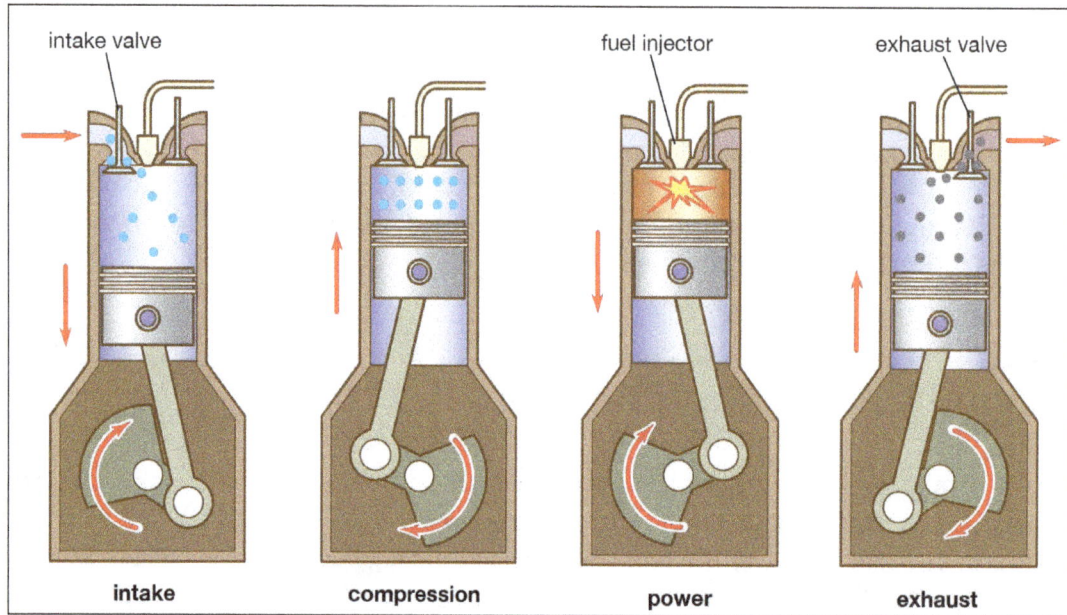

Four-stroke engine cycle.

The intake stroke is succeeded by the *compression* stroke which effectively starts at the intake valve closure. Its purpose is to prepare the mixture for combustion by increasing its temperature and pressure. Combustion is initiated by the energy released through the spark plug towards the end of the compression stroke and is associated with a rapid rise in the cylinder pressure.

The *power or expansion* stroke starts with the piston at TDC of compression and ends at BDC. At this point, the high temperature and pressure gases generated during combustion push the piston down, thus forcing the crank to rotate. Just before the piston reaches BDC, the exhaust valve(s) opens and the burned gases are allowed to exit the cylinder due to the differential pressure between the cylinder and the exhaust manifold.

This *exhaust* stroke completes the engine cycle by evacuating the cylinder from burned, partially-burned or even unburned gases escaping the combustion process; the next engine cycle starts when the intake valve opens near TDC and the exhaust valve closes a few degrees crank angle later.

It is important to note that the properties of gasoline, in association with combustion chamber geometry, exert a significant influence on combustion duration, rate of pressure rise and *pollutant formation*. Under certain conditions, the mixture at the end gas may autoignite before the flame reaches that part of the cylinder, leading to *knock* which gives rise to high-intensity and frequency pressure oscillations.

The tendency of gasoline fuel to resist *autoignition* and thus prevent possible damage to the engine as a result of *knock* is characterised by its *octane number*. Until recently, the addition of a small quantity of lead into the gasoline was the preferred method for suppressing knock but the associated health risks, combined with the need to use catalysts for reducing exhaust emissions, have necessitated the introduction of unleaded gasoline. This requires a reduction of the engine's compression ratio (ratio of the cylinder volume at BDC to the volume at TDC) in order to prevent knock with undesirable effects on thermal efficiency.

As already mentioned, the four-stroke cycle, also known as Otto cycle after its inventor Nicolaus Otto who built the first engine in 1876, produces a power stroke for every two crankshaft revolutions. One way to increase the power output of a given engine size is to convert it to a two stroke cycle in which power is produced during every engine revolution.

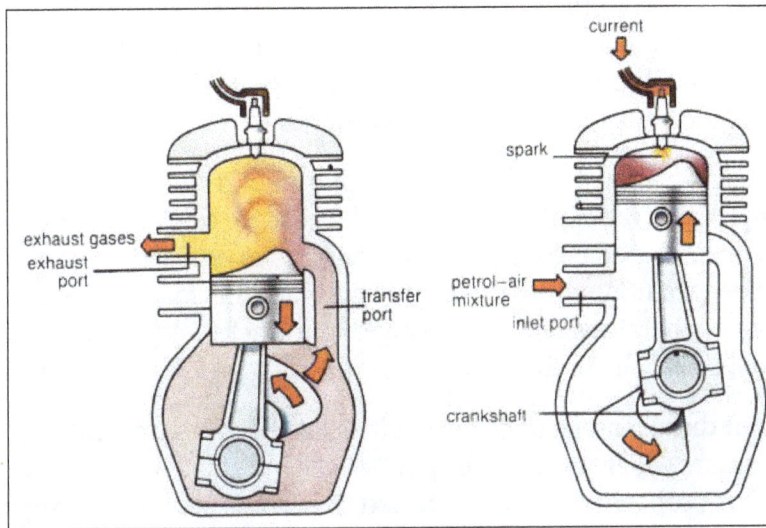

Two-stroke engine cycle.

Because this mode of operation gives rise to increased power output—albeit not to the double levels expected from simple calculations—it has been extensively used in motorcycle, passenger car and marine applications with both spark-ignition and diesel engines. An additional advantage is the simple design of two-stroke engines since they can operate with side ports in the liner, covered and uncovered by piston motion, instead of the bulky and complicated overhead cam arrangement.

In the two-stroke cycle, the *compression* stroke starts after the inlet and exhaust side ports are covered by the piston; the fuel/air mixture is compressed and then ignited by a spark-plug, similar to ignition in a four-stroke gasoline engine, to initiate combustion near TDC. At the same time, fresh charge is allowed to enter the crankcase before its subsequent compression by the downward-moving piston during the *power or expansion* stroke. During this period, burned gases push the piston until it reaches BDC, which allows first the exhaust ports and then the intake (transfer) ports to be uncovered. The opening of the exhaust ports permits the burned gases to exit the cylinder while partly at the same time the fresh charge, which has been compressed in the crankcase, enters the cylinder through the properly orientated transfer ports.

The overlapping of the induction and exhaust strokes in two-stroke cycle engines is responsible for some of the fresh charge flowing directly out of the cylinder during the scavenging process. Despite various attempts to reduce the magnitude of this problem by introducing a deflector into the piston and directing the incoming charge away from the location of the exhaust ports, charging efficiency in conventional two-stroke engines remains relatively low. A solution to this problem is to introduce the fuel directly into the cylinder, separately from the fresh air, through air-assisted injectors during the period when both the exhaust and transfer ports are closed. Despite the short period available for mixing, air-assisted atomizers can achieve a homogeneous lean mixture at the time of ignition by generating gasoline droplets of less than 40 μm mean diameter, which vaporize very easily during the compression stroke.

Amongst the various types of internal combustion engines, the diesel or compression-ignition engine is renowned for its high efficiency, reduced fuel consumption and relatively low total gaseous emissions. Its name comes from the German engineer Rudolf Diesel who in 1892 described in his patent a form of internal combustion engine which does not require an external source of ignition and where combustion is initiated by the auto ignition of the liquid fuel injected into the high temperature and pressure air towards the end of the compression stroke.

The inherent efficiency advantages of the diesel engine stem from its lean overall mixture ratios, the high engine compression ratios afforded due to the absence of end-gas ignition (knock) and the greater expansion ratios. As a consequence, diesel engines in either the two-stroke or four-stroke configuration have been traditionally the preferred power plants for commercial applications such as ships/boats, energy-generating sets, locomotives and tracks and, over the last 20 years or so, passenger cars as well especially in Europe.

The low power-output disadvantage of diesel engines has been circumvented by the use of superchargers or turbochargers which increase the power/weight ratio of an engine through an increase of the inlet air density. Turbochargers are expected to become standard components of all future diesel engines, irrespective of application.

The operation of the diesel engine differs from that of the spark-ignition engine mainly in the way the mixture is formed prior to combustion. Only air is inducted into the engine through a helical or directed port and the fuel mixes with air during the compression stroke, following its injection at high pressure into a prechamber indirect-injection diesel or IDI) or into the main chamber (direct-injection diesel or DI) just, before combustion is started.

The need to achieve good fuel/air mixing in diesel engines is satisfied by high-pressure fuel injection systems which generate droplets of about 40 μm mean diameter. For passenger cars, the fuel injection systems consist of a rotary pump, delivery pipes and fuel injector nozzles which vary in their design according to the application; direct-injection diesel engines use hole-type nozzles while indirect-injection diesels employ pintle-type nozzles. Larger diesel engines use in-line fuel-injection pumps, unit injectors (pump and nozzle combined in one unit) or individual single-barrel pumps which are mounted close to each cylinder.

Over the last 20 years or so, the realization that the resources of crude oil are finite and that the environment we live in is becoming more and more polluted, has urged governments to introduce laws which limit the *exhaust emission levels* of vehicles and engines of all types. Since their introduction in Japan and the USA in the late 60s and in Europe in 1970, emission regulations are consistently becoming more stringent and engine manufacturers are facing their toughest ever challenge with the standards agreed for 1996 onwards, it is uncertain whether existing engines will satisfy these limits despite the desperate attempts by engineers worldwide.

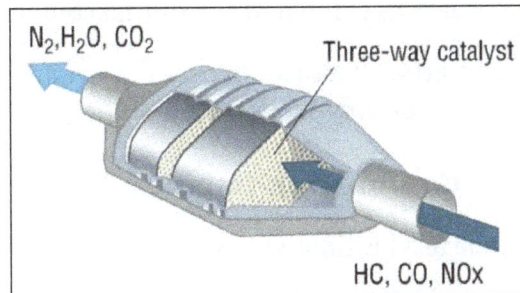

Model of a three-way catalytic converter.

The major pollutants in spark-ignition engines are hydrocarbons (HC), carbon monoxide (CO) and oxides of nitrogen ($NO_x = NO + NO_2$) while in diesel engines, NO_x and particulates—which consist of soot particles formed during combustion of lubricating oil and hydrocarbons—are the most harmful.

At present three-way catalysts, which are a standard component of today's passenger cars equipped with spark-ignition engine running on unleaded gasoline, allow about 90% reduction of the emitted HC, CO and NO_x by converting them into carbon dioxide (CO_2), water (H_2O) and N_2.

Unfortunately these catalysts require stoichiometric (air-fuel ratio of ~14.5) engine operation, which is undesirable from both the fuel consumption and CO_2 emissions points of view. An alternative approach is the lean burn concept which offers promise for simultaneous reduction of fuel consumption and exhaust emissions through satisfactory combustion of lean mixtures with much higher than 20 air-fuel ratios. It is expected that the development of lean burn catalysts with conversion efficiencies over 60% may allow lean burn engines to satisfy future emissions legislation; this is an area of active research in both industry and academia. On the other hand, new diesel engines depend on two-way or oxidizing catalysts for reduction of exhaust particulates through conversion of HC into CO_2 and H_2O, and on exhaust gas recirculation and retarded injection timing for reduction of NO_x levels.

Advantages of Internal Combustion Engines

- Size of engine is very less compared to external combustion engines.

- Power to weight ratio is high.

- Very suitable for small power requirement applications.

- Usually more portable than their counterpart external combustion engines.

- Safer to operate.

- Starting time is very less.

- High efficiency than external combustion engine.

- No chances of leakage of working fluids.

- Requires less maintenance.

- Lubricant consumption is less as compared to external combustion engines.

- In case of reciprocating internal combustion overall working temperature is low because. peak temperature is reached for only small period of time (only at detonation of fuel).

Disadvantages of Internal Combustion Engines

- Variety of fuels that can be used is limited to very fine quality gaseous and liquid fuel.

- Fuel used is very costly like gasoline or diesel.

- Engine emissions are generally high compared to external combustion engine.

- Not suitable of large scale power generation.

- In case of reciprocating internal combustion noise is generated due to detonation of fuel.

Types and Applications of Internal Combustion Engines

- Gasoline Engines: Automotive, Marine, Aircraft.

- Gas Engines: Industrial Power.

- Diesel Engines: Automotive, Railways, Power, Marine.

- Gas Turbines: Power, Aircraft, Industrial, Marine.

2
Classification of Internal Combustion Engine

The internal combustion engine is broadly classified into a number of categories, namely, reciprocating engine, diesel engine, spark-ignition engine, jet engine, gas turbine and rotatory engine. The topics elaborated in this chapter will help in gaining a better perspective about these categories of internal combustion engine.

Reciprocating Engine

Reciprocating engines and turboprop engines work in combination with a propeller to produce thrust. Turbojet and turbofan engines produce thrust by increasing the velocity of air flowing through the engine. All of these powerplants also drive the various systems that support the operation of an aircraft.

Most small aircraft are designed with reciprocating engines. The name is derived from the back-and-forth, or reciprocating, movement of the pistons that produces the mechanical energy necessary to accomplish work.

Driven by a revitalization of the general aviation (GA) industry and advances in both material and engine design, reciprocating engine technology has improved dramatically over the past two decades. The integration of computerized engine management systems has improved fuel efficiency, decreased emissions, and reduced pilot workload.

Reciprocating engines operate on the basic principle of converting chemical energy (fuel) into mechanical energy. This conversion occurs within the cylinders of the engine through the process of combustion. The two primary reciprocating engine designs are the spark ignition and the compression ignition. The spark ignition reciprocating engine has served as the powerplant of choice for many years. In an effort to reduce operating costs, simplify design, and improve reliability, several engine manufacturers are turning to compression ignition as a viable alternative. Often referred to as jet fuel piston engines, compression ignition engines have the added advantage of utilizing readily available and lower cost diesel or jet fuel.

The main mechanical components of the spark ignition and the compression ignition engine are essentially the same. Both use cylindrical combustion chambers and pistons that travel the length of the cylinders to convert linear motion into the rotary motion of the crankshaft. The main difference between spark ignition and compression ignition is the process of igniting the fuel. Spark ignition engines use a spark plug to ignite a pre-mixed fuel-air mixture. (Fuel-air mixture is the ratio of the "weight" of fuel to the "weight" of air in the mixture to be burned.) A compression ignition engine first compresses the air in the cylinder, raising its temperature to a degree necessary for automatic ignition when fuel is injected into the cylinder.

These two engine designs can be further classified as:

- Cylinder arrangement with respect to the crankshaft: Radial, in-line, v-type, or opposed.

- Operating cycle: Two or four.

- Method of cooling: Liquid or air.

Radial engines were widely used during World War II and many are still in service today. With these engines, a row or rows of cylinders are arranged in a circular pattern around the crankcase. The main advantage of a radial engine is the favorable power-to-weight ratio.

Radial engine.

In-line engines have a comparatively small frontal area, but their power-to-weight ratios are relatively low. In addition, the rearmost cylinders of an air-cooled, in-line engine receive very little cooling air, so these engines are normally limited to four or six cylinders. V-type engines provide more horsepower than in-line engines and still retain a small frontal area.

Continued improvements in engine design led to the development of the horizontally-opposed engine, which remains the most popular reciprocating engines used on smaller aircraft. These engines always have an even number of cylinders, since a cylinder on one side of the crankcase "opposes" a cylinder on the other side. The majority of these engines are air cooled and usually are mounted in a horizontal position when installed on fixed-wing airplanes. Opposed-type engines have high power-to-weight ratios because they have a comparatively small, lightweight crankcase. In addition, the compact cylinder arrangement reduces the engine's frontal area and allows a streamlined installation that minimizes aerodynamic drag.

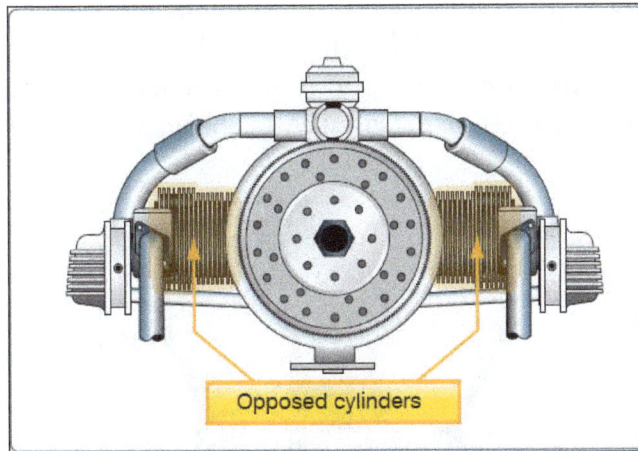

Horizontally opposed engine.

Depending on the engine manufacturer, all of these arrangements can be designed to utilize spark or compression ignition and operate on either a two- or four-stroke cycle.

In a two-stroke engine, the conversion of chemical energy into mechanical energy occurs over a two-stroke operating cycle. The intake, compression, power, and exhaust processes occur in only two strokes of the piston rather than the more common four strokes. Because a two-stroke engine has a power stroke upon each revolution of the crankshaft, it typically has higher power-to-weight ratio than a comparable four-stroke engine. Due to the inherent inefficiency and disproportionate emissions of the earliest designs, use of the two-stroke engine has been limited in aviation.

Recent advances in material and engine design have reduced many of the negative characteristics associated with two-stroke engines. Modern two-stroke engines often use conventional oil sumps, oil pumps, and full pressure fed lubrication systems. The use of direct fuel injection and pressurized air, characteristic of advanced compression ignition engines, make two-stroke compression ignition engines a viable alternative to the more common four-stroke spark ignition designs.

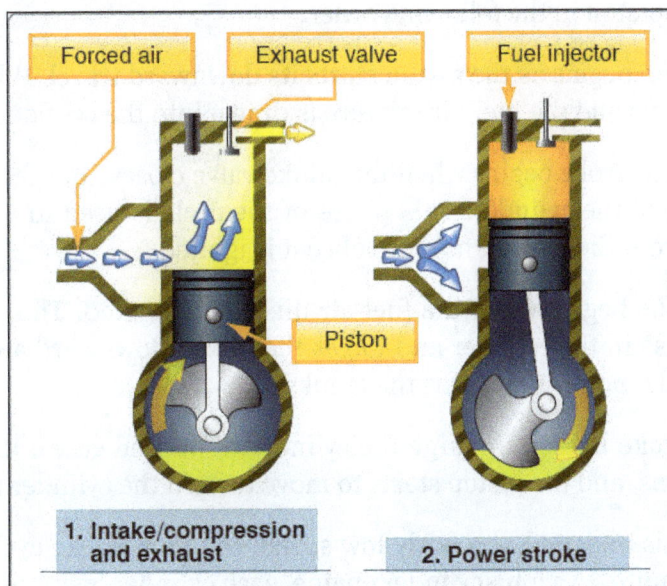

Two-stroke compression ignition.

Spark ignition four-stroke engines remain the most common design used in GA today. The main parts of a spark ignition reciprocating engine include the cylinders, crankcase, and accessory housing. The intake/exhaust valves, spark plugs, and pistons are located in the cylinders. The crankshaft and connecting rods are located in the crankcase. The magnetos are normally located on the engine accessory housing.

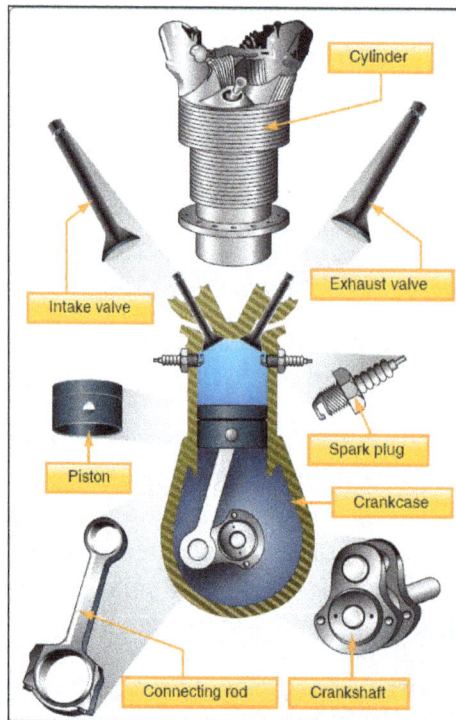

Main components of a spark ignition reciprocating engine.

In a four-stroke engine, the conversion of chemical energy into mechanical energy occurs over a four-stroke operating cycle. The intake, compression, power, and exhaust processes occur in four separate strokes of the piston in the following order:

- The intake stroke begins as the piston starts its downward travel. When this happens, the intake valve opens and the fuel-air mixture is drawn into the cylinder.

- The compression stroke begins when the intake valve closes, and the piston starts moving back to the top of the cylinder. This phase of the cycle is used to obtain a much greater power output from the fuel-air mixture once it is ignited.

- The power stroke begins when the fuel-air mixture is ignited. This causes a tremendous pressure increase in the cylinder and forces the piston downward away from the cylinder head, creating the power that turns the crankshaft.

- The exhaust stroke is used to purge the cylinder of burned gases. It begins when the exhaust valve opens, and the piston starts to move toward the cylinder head once again.

Even when the engine is operated at a fairly low speed, the four-stroke cycle takes place several hundred times each minute. In a four-cylinder engine, each cylinder operates on a different stroke. Continuous rotation of a crankshaft is maintained by the precise timing of the power strokes in each cylinder. Continuous operation of the engine depends on the simultaneous function of auxiliary systems, including the induction, ignition, fuel, oil, cooling, and exhaust systems.

Two-stroke Engine

A two-stroke (or two-cycle) engine is a type of internal combustion engine which completes a power cycle with two strokes (up and down movements) of the piston during only one crankshaft revolution. This is in contrast to a "four-stroke engine", which requires four strokes of the piston to complete a power cycle during two crankshaft revolutions. In a two-stroke engine, the end of the combustion stroke and the beginning of the compression stroke happen simultaneously, with the intake and exhaust (or scavenging) functions occurring at the same time.

Two-stroke engines often have a high power-to-weight ratio, power being available in a narrow range of rotational speeds called the "power band". Compared to four-stroke engines, two-stroke engines have a greatly reduced number of moving parts, and so can be more compact and significantly lighter.

Emissions

Crankcase-compression two-stroke engines, such as common small gasoline-powered engines, are lubricated by a petroil mixture in a total-loss system. Oil is mixed in with their petrol fuel beforehand, in a ratio of around 1:50. All that oil then forms emissions, either by being burned in the engine or as oily droplets in the exhaust. This creates more exhaust emissions, particularly hydrocarbons, than four-stroke engines of comparable power output. The combined opening time of the intake and exhaust ports in some 2-stroke designs can also allow some amount of unburned fuel vapors to exit in the exhaust stream. The high combustion temperatures of small air-cooled engines may also give high NO_x emissions.

Applications

Two-stroke petrol engines are preferred when mechanical simplicity, light weight, and high power-to-weight ratio are design priorities. With the traditional lubrication technique of mixing oil into the fuel, they also have the advantage of working in any orientation, as there is no oil reservoir dependent on gravity; this is an essential property for hand-held power tools such as chainsaws.

1966 Saab Sport.

Lateral view of a two-stroke Forty series British Seagull outboard engine.

A number of mainstream automobile manufacturers have used two-stroke engines in the past, including the Swedish Saab and German manufacturers DKW, Auto-Union, VEB Sachsenring Automobilwerke Zwickau, and VEB Automobilwerk Eisenach. The Japanese manufacturers Suzuki and Subaru did the same in the 1970s. Production of two-stroke cars ended in the 1980s in the West, due to increasingly stringent regulation of air pollution. Eastern Bloc countries continued until around 1991, with the Trabant and Wartburg in East Germany. Two-stroke engines are still found in a variety of small propulsion applications, such as outboard motors, high-performance small-capacity motorcycles, mopeds, and dirt bikes, underbones, scooters, tuk-tuks, snowmobiles, karts, ultralight airplanes, model airplanes and other model vehicles. They are also common in power tools used outdoors, such as lawn mowers, chainsaws, and weed-wackers.

With direct fuel injection and a sump-based lubrication system, a two-stroke engine produces air pollution no worse than a four-stroke, and it can achieve higher thermodynamic efficiency. Therefore, the cycle has historically also been used in large diesel engines, mostly large industrial and marine engines, as well as some trucks and heavy machinery. There are several experimental designs intended for automobile use: For instance, Lotus of Norfolk, UK, had in 2008 a prototype direct-injection two-stroke engine intended for alcohol fuels called the Omnivore which it is demonstrating in a version of the Exige.

Different Two-stroke Design Types

Although, the principles remain the same, the mechanical details of various two-stroke engines differ depending on the type. The design types vary according to the method of introducing the charge to the cylinder, the method of scavenging the cylinder (exchanging burnt exhaust for fresh mixture) and the method of exhausting the cylinder.

Piston-controlled Inlet Port

Piston port is the simplest of the designs and the most common in small two-stroke engines. All functions are controlled solely by the piston covering and uncovering the ports as it moves up and down in the cylinder. In the 1970s, Yamaha worked out some basic principles for this system.

They found that, in general, widening an exhaust port increases the power by the same amount as raising the port, but the power band does not narrow as it does when the port is raised. However, there is a mechanical limit to the width of a single exhaust port, at about 62% of the bore diameter for reasonable ring life. Beyond this, the rings will bulge into the exhaust port and wear quickly. A maximum 70% of bore width is possible in racing engines, where rings are changed every few races. Intake duration is between 120 and 160 degrees. Transfer port time is set at a minimum of 26 degrees. The strong low pressure pulse of a racing two-stroke expansion chamber can drop the pressure to -7 PSI when the piston is at bottom dead center, and the transfer ports nearly wide open. One of the reasons for high fuel consumption in two-strokes is that some of the incoming pressurized fuel-air mixture is forced across the top of the piston, where it has a cooling action, and straight out the exhaust pipe. An expansion chamber with a strong reverse pulse will stop this out-going flow. A fundamental difference from typical four-stroke engines is that the two-stroke's crankcase is sealed and forms part of the induction process in gasoline and hot bulb engines. Diesel two strokes often add a Roots blower or piston pump for scavenging.

Reed Inlet Valve

A Cox Babe Bee 0.049 cubic inch (0.8 cubic cm) reed valve engine, disassembled, uses glow plug ignition. The mass is 64 grams.

The reed valve is a simple but highly effective form of check valve commonly fitted in the intake tract of the piston-controlled port. They allow asymmetric intake of the fuel charge, improving power and economy, while widening the power band. They are widely used in motorcycle, ATV and marine outboard engines.

Rotary Inlet Valve

The intake pathway is opened and closed by a rotating member. A familiar type sometimes seen on small motorcycles is a slotted disk attached to the crankshaft which covers and uncovers an opening in the end of the crankcase, allowing charge to enter during one portion of the cycle (aka disc valve).

Another form of rotary inlet valve used on two-stroke engines employs two cylindrical members with suitable cutouts arranged to rotate one within the other - the inlet pipe having passage to the crankcase only when the two cutouts coincide. The crankshaft itself may form one of the members, as in most glow plug model engines. In another embodiment, the crank disc is arranged to be a close-clearance fit in the crankcase, and is provided with a cutout which lines up with an inlet passage in the crankcase wall at the appropriate time, as in Vespa motor scooters.

The advantage of a rotary valve is that it enables the two-stroke engine's intake timing to be asymmetrical, which is not possible with piston-port type engines. The piston-port type engine's intake timing opens and closes before and after top dead center at the same crank angle, making it symmetrical, whereas the rotary valve allows the opening to begin and close earlier.

Rotary valve engines can be tailored to deliver power over a wider speed range or higher power over a narrower speed range than either piston port or reed valve engine. Where a portion of the rotary valve is a portion of the crankcase itself, it is particularly important that no wear is allowed to take place.

Cross-flow-scavenged

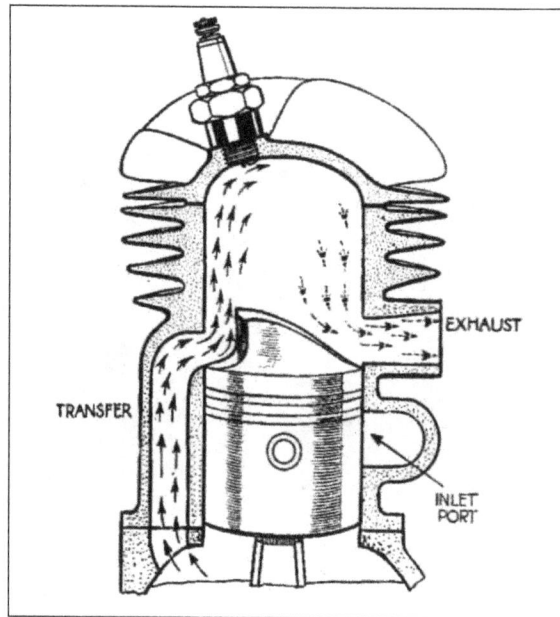

Deflector piston with cross-flow scavenging.

In a cross-flow engine, the transfer and exhaust ports are on opposite sides of the cylinder, and a deflector on the top of the piston directs the fresh intake charge into the upper part of the cylinder, pushing the residual exhaust gas down the other side of the deflector and out the exhaust port. The deflector increases the piston's weight and exposed surface area, affecting piston cooling and also making it difficult to achieve an efficient combustion chamber shape. This design has been superseded since the 1960s by the loop scavenging method (below), especially for motorbikes, although for smaller or slower engines, such as lawn mowers, the cross-flow-scavenged design can be an acceptable approach.

Loop-scavenged

This method of scavenging uses carefully shaped and positioned transfer ports to direct the flow of fresh mixture toward the combustion chamber as it enters the cylinder. The fuel/air mixture strikes the cylinder head, then follows the curvature of the combustion chamber, and then is deflected downward.

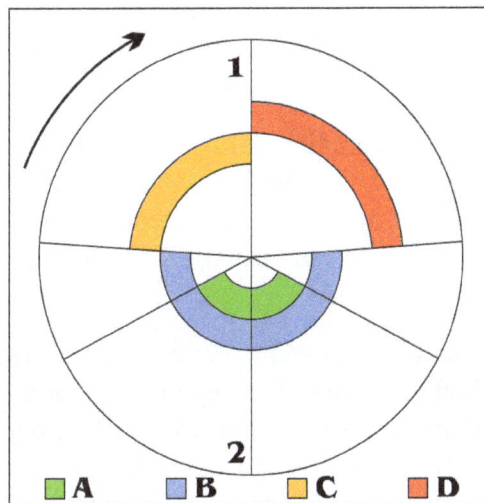

In figure, the two-stroke cycle:

1. Top dead center (TDC),

2. Bottom dead center (BDC),

    A: Intake/scavenging,

    B: Exhaust,

    C: Compression,

    D: Expansion (power).

This not only prevents the fuel/air mixture from traveling directly out the exhaust port, but also creates a swirling turbulence which improves combustion efficiency, power and economy. Usually, a piston deflector is not required, so this approach has a distinct advantage over the cross-flow scheme.

Often referred to as "Schnuerle" (or "Schnürle") loop scavenging after Adolf Schnürle, the German inventor of an early form in the mid-1920s, it became widely adopted in that country during the 1930s and spread further afield after World War II.

Loop scavenging is the most common type of fuel/air mixture transfer used on modern two-stroke engines. Suzuki was one of the first manufacturers outside of Europe to adopt loop-scavenged two-stroke engines. This operational feature was used in conjunction with the expansion chamber exhaust developed by German motorcycle manufacturer, MZ and Walter Kaaden.

Loop scavenging, disc valves and expansion chambers worked in a highly coordinated way to significantly increase the power output of two-stroke engines, particularly from the Japanese manufacturers Suzuki, Yamaha and Kawasaki. Suzuki and Yamaha enjoyed success in grand Prix motorcycle racing in the 1960s due in no small way to the increased power afforded by loop scavenging.

An additional benefit of loop scavenging was the piston could be made nearly flat or slightly dome shaped, which allowed the piston to be appreciably lighter and stronger, and consequently to tolerate higher engine speeds. The "flat top" piston also has better thermal properties and is less prone to uneven heating, expansion, piston seizures, dimensional changes and compression losses.

SAAB built 750 and 850 cc 3-cylinder engines based on a DKW design that proved reasonably successful employing loop charging. The original SAAB 92 had a two-cylinder engine of comparatively low efficiency. At cruising speed, reflected wave exhaust port blocking occurred at too low a frequency. Using the asymmetric three-port exhaust manifold employed in the identical DKW engine improved fuel economy.

The 750 cc standard engine produced 36 to 42 hp, depending on the model year. The Monte Carlo Rally variant, 750 cc (with a filled crankshaft for higher base compression), generated 65 hp. An 850 cc version was available in the 1966 SAAB Sport (a standard trim model in comparison to the deluxe trim of the Monte Carlo). Base compression comprises a portion of the overall compression ratio of a two-stroke engine. Work published at SAE in 2012 points that loop scavenging is under every circumstance more efficient than cross-flow scavenging.

Uniflow-scavenged

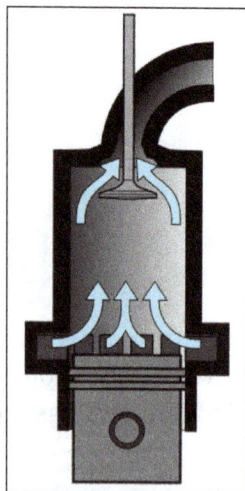

Uniflow scavenging.

In a uniflow engine, the mixture, or "charge air" in the case of a diesel, enters at one end of the cylinder controlled by the piston and the exhaust exits at the other end controlled by an exhaust valve or piston. The scavenging gas-flow is therefore in one direction only, hence the name uniflow. The valved arrangement is common in on-road, off-road and stationary two-stroke engines (Detroit Diesel), certain small marine two-stroke engines (Gray Marine), certain railroad two-stroke diesel locomotives (Electro-Motive Diesel) and large marine two-stroke main propulsion engines (Wärtsilä). Ported types are represented by the opposed piston design in which there are two pistons in each cylinder, working in opposite directions such as the Junkers Jumo 205 and Napier Deltic. The once-popular split-single design falls into this class, being effectively a folded uniflow. With advanced angle exhaust timing, uniflow engines can be supercharged with a crankshaft-driven (piston or Roots) blower.

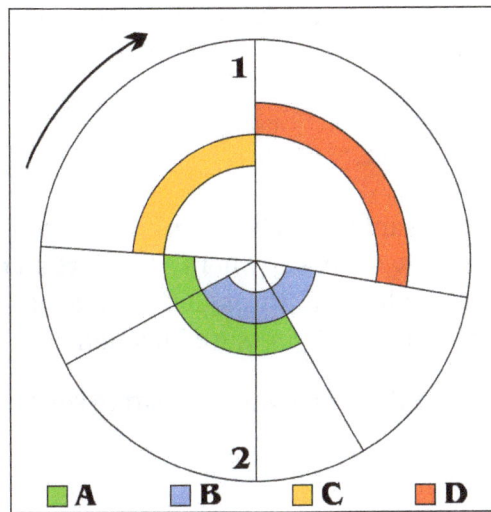

In figure, the uniflow two-stroke cycle.

1. Top dead center (TDC)

2. Bottom dead center (BDC)

   A: Intake (effective scavenging, 135°–225°; necessarily symmetric about BDC; Diesel injection is usually initiated at 4° before TDC)

   B: Exhaust

   C: Compression

   D: Expansion (power)

Stepped Piston Engine

The piston of this engine is "top-hat" shaped; the upper section forms the regular cylinder, and the lower section performs a scavenging function. The units run in pairs, with the lower half of one piston charging an adjacent combustion chamber.

This system is still partially dependent on total loss lubrication (for the upper part of the piston), the other parts being sump lubricated with cleanliness and reliability benefits. The piston weight is only about 20% heavier than a loop-scavenged piston because skirt thicknesses can be less. Bernard Hooper Engineering Ltd. (BHE) is one of the more recent engine developers using this approach.

Power Valve Systems

Many modern two-stroke engines employ a power valve system. The valves are normally in or around the exhaust ports. They work in one of two ways: either they alter the exhaust port by closing off the top part of the port, which alters port timing, such as Rotax R.A.V.E, Yamaha YPVS, Honda RC-Valve, Kawasaki K.I.P.S., Cagiva C.T.S. or Suzuki AETC systems, or by altering the volume of the exhaust, which changes the resonant frequency of the expansion chamber, such as the Suzuki SAEC and Honda V-TACS system. The result is an engine with better low-speed power without sacrificing high-speed power. However, as power valves are in the hot gas flow they need regular maintenance to perform well.

Direct Injection

Direct injection has considerable advantages in two-stroke engines, eliminating some of the waste and pollution caused by carbureted two-strokes where a proportion of the fuel/air mixture entering the cylinder goes directly out, unburned, through the exhaust port. Two systems are in use, low-pressure air-assisted injection, and high pressure injection.

Since the fuel does not pass through the crankcase, a separate source of lubrication is needed.

Diesel

Brons two-stroke V8 Diesel engine driving an N.V. Heemaf generator.

Diesel engines rely solely on the heat of compression for ignition. In the case of Schnuerle ported and loop-scavenged engines, intake and exhaust happens via piston-controlled ports. A uniflow diesel engine takes in air via scavenge ports, and exhaust gases exit through an overhead poppet valve. Two-stroke diesels are all scavenged by forced induction. Some designs use a mechanically driven Roots blower, whilst marine diesel engines normally use exhaust-driven turbochargers, with electrically driven auxiliary blowers for low-speed operation when exhaust turbochargers are unable to deliver enough air.

Marine two-stroke diesel engines directly coupled to the propeller are able to start and run in either direction as required. The fuel injection and valve timing is mechanically readjusted by using a different set of cams on the camshaft. Thus, the engine can be run in reverse to move the vessel backwards.

Lubrication

Most small petrol two-stroke engines cannot be lubricated by oil contained in their crankcase and sump, since the crankcase is being used to pump fuel-air mixture into the cylinder. Over a short period, the constant stream of fuel-air mixture would carry away the lubricating oil into the combustion chamber while thinning the remainder with condensing petrol. Traditionally, the moving parts (both rotating crankshaft and sliding piston) were instead lubricated by a premixed fuel-oil mixture (at a ratio between 16:1 and 100:1). As late as the 1970s, petrol stations would often have a separate pump to deliver such a premix fuel to motorcycles. Even then, in many cases, the rider would carry a bottle of their own two-stroke oil.

Two-stroke oils which became available worldwide in the 1970s are specifically designed to mix with petrol and be burnt in the combustion chamber without leaving undue unburnt oil or ash. This led to a marked reduction in spark plug fouling, which had previously been a factor in two-stroke engines.

More recent two-stroke engines might pump lubrication from a separate tank of two-stroke oil. The supply of this oil is controlled by the throttle position and engine speed. Examples are found in Yamaha's PW80 (Pee-wee), a small, 80cc two-stroke dirt bike designed for young children, and many two-stroke snowmobiles. The technology is referred to as auto-lube. This is still a total-loss system with the oil being burnt the same as in the pre-mix system; however, given that the oil is not properly mixed with the fuel when burned in the combustion chamber, it translates into a slightly more efficient lubrication. This lubrication method also pays dividends in terms of user friendliness by eliminating the user's need to mix the gasoline at every refill, makes the motor much less susceptible to atmospheric conditions (Ambient temperature, elevation) and ensures proper engine lubrication, with less oil at light loads (such as idle) and more oil at high loads (such as full throttle). Some companies, such as Bombardier, had some oil pump designs have no oil injected at idle to reduce smoke levels, as the loading on the engine parts was light enough to not require additional lubrication beyond the low levels that the fuel provides. Ultimately oil injection is still the same as premixed gasoline in that the oil is burnt in the combustion chamber (albeit not as completely as pre-mix) and the gas is still mixed with the oil, although not as thoroughly as in pre-mix. In addition, this method requires extra mechanical parts to pump the oil from the separate tank, to the carburetor or throttle body. In applications where performance, simplicity and/or dry weight are significant considerations, the pre-mix lubrication method is almost always used. For example, a two-stroke engine in a motocross bike pays major consideration to performance, simplicity and weight. Chainsaws and brush cutters must be as light as possible to reduce user fatigue and hazard, especially when used in a professional work environment.

All two-stroke engines running on a petrol/oil mix will suffer oil starvation if forced to rotate at speed with the throttle closed, e.g. motorcycles descending long hills and perhaps when decelerating gradually from high speed by changing down through the gears. Two-stroke cars (such as those that were popular in Eastern Europe in the mid-20th century) were in particular danger and were usually fitted with freewheel mechanisms in the powertrain, allowing the engine to idle when the throttle was closed, requiring the use of the brakes in all slowing situations.

Large two-stroke engines, including diesels, normally use a sump lubrication system similar to four-stroke engines. The cylinder must still be pressurized, but this is not done from the crankcase, but by an ancillary Roots-type blower or a specialized turbocharger (usually a turbo-compressor system) which has a "locked" compressor for starting (and during which it is powered by the engine's crankshaft), but which is "unlocked" for running (and during which it is powered by the engine's exhaust gases flowing through the turbine).

Two-stroke Reversibility

For the purpose of this discussion, it is convenient to think in motorcycle terms, where the exhaust pipe faces into the cooling air stream, and the crankshaft commonly spins in the same axis and direction as do the wheels i.e. "forward". Some of the considerations discussed here apply to four-stroke engines (which cannot reverse their direction of rotation without considerable modification), almost all of which spin forward, too.

Regular gasoline two-stroke engines will run backwards for short periods and under light load with little problem, and this has been used to provide a reversing facility in microcars, such as the Messerschmitt KR200, that lacked reverse gearing. Where the vehicle has electric starting, the motor will be turned off and restarted backwards by turning the key in the opposite direction. Two-stroke golf carts have used a similar kind of system. Traditional flywheel magnetos (using contact-breaker points, but no external coil) worked equally well in reverse because the cam controlling the points is symmetrical, breaking contact before top dead center (TDC) equally well whether running forwards or backwards. Reed-valve engines will run backwards just as well as piston-controlled porting, though rotary valve engines have asymmetrical inlet timing and will not run very well.

There are serious disadvantages to running many engines backwards under load for any length of time, and some of these reasons are general, applying equally to both two-stroke and four-stroke engines. This disadvantage is accepted in most cases where cost, weight and size are major considerations. The problem comes about because in "forwards" running the major thrust face of the piston is on the back face of the cylinder which, in a two-stroke particularly, is the coolest and best-lubricated part. The forward face of the piston in a trunk engine is less well-suited to be the major thrust face since it covers and uncovers the exhaust port in the cylinder, the hottest part of the engine, where piston lubrication is at its most marginal. The front face of the piston is also more vulnerable since the exhaust port, the largest in the engine, is in the front wall of the cylinder. Piston skirts and rings risk being extruded into this port, so it is always better to have them pressing hardest on the opposite wall (where there are only the transfer ports in a crossflow engine) and there is good support. In some engines, the small end is offset to reduce thrust in the intended rotational direction and the forward face of the piston has been made thinner and lighter to compensate; but when running backwards, this weaker forward face suffers increased mechanical stress it was not designed to resist. This can be avoided by the use of crossheads and also using thrust bearings to isolate the engine from end loads.

Large two-stroke ship diesels are sometimes made to be reversible. Like four-stroke ship engines (some of which are also reversible) they use mechanically operated valves, so require additional camshaft mechanisms. These engines use crossheads to eliminate sidethrust on the piston and isolate the under-piston space from the crankcase.

On top of other considerations, the oil-pump of a modern two-stroke may not work in reverse, in which case the engine will suffer oil starvation within a short time. Running a motorcycle engine backwards is relatively easy to initiate, and in rare cases, can be triggered by a back-fire. It is not advisable.

Model airplane engines with reed-valves can be mounted in either tractor or pusher configuration without needing to change the propeller. These motors are compression ignition, so there are no ignition timing issues and little difference between running forward and running backward.

Schnuerle Porting

Schnuerle porting is a system to improve efficiency of a valveless two-stroke engine by giving better scavenging. The intake and exhaust ports cut in the cylinder wall are shaped to give a more efficient transfer of intake and exhaust gases.

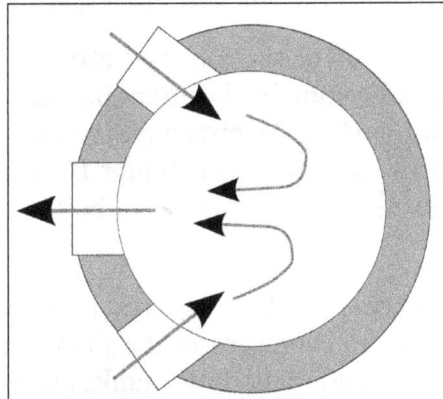

Cylinder and ports, viewed from above.

Gas flow within the two-stroke engine is even more critical than for a four-stroke engine, as the two flows are both entering and leaving the combustion chamber simultaneously. A well-defined flow pattern is required, avoiding any turbulent mixing. The efficiency of the two-stroke engine depends on effective scavenging, the more complete replacement of the old spent charge with a fresh charge.

Apart from large diesels with separate superchargers, two-stroke engines are generally piston-ported and use their crankcase beneath the piston for compression. The cylinder has a transfer port (inlet from crankcase to cylinder) and an exhaust port cut into it. These are opened, as the piston moves downwards past them; with the higher exhaust port opening earlier as the piston descends; and closing later as the piston rises.

The simplest arrangement is a single transfer and single exhaust port, opposite each other. This "cross scavenging" performs poorly, as there is tendency for the flow to pass from the inlet directly to the exhaust, wasting some of the fuel mixture and also poorly scavenging the upper part of the chamber. Before Schnuerle porting, a deflector on top of the piston was used to direct the gas flow from the transfer port upwards, in a U-shaped loop around the combustion chamber roof and then down and out through the exhaust port. Apart from the gas flow never quite following this ideal path and tending to mix instead, this also gave a poorly shaped combustion chamber with long, thin flame paths.

In 1926, the German engineer Adolf Schnürle developed the system of ports that bears his name. The ports were relocated to both be on the same side of the cylinder, with the transfer port being split into two angled ports, one on either side of the exhaust port. A deflector piston was no longer required. The gas flow was now a circular loop, flowing in and across the piston crown from the transfer ports, up and around the combustion chamber and then out through the exhaust port.

With Schnuerle porting, the piston crown may be of any shape, even bowl shaped. This permits a far better combustion chamber shape and flame path, giving better combustion, particularly at high speeds.

Four-stroke Engine

A four-stroke (also four-cycle) engine is an internal combustion (IC) engine in which the piston completes four separate strokes while turning the crankshaft. A stroke refers to the full travel of the piston along the cylinder, in either direction. The four separate strokes are termed:

- Intake: Also known as induction or suction. This stroke of the piston begins at top dead center (T.D.C.) and ends at bottom dead center (B.D.C.). In this stroke the intake valve must be in the open position while the piston pulls an air-fuel mixture into the cylinder by producing vacuum pressure into the cylinder through its downward motion. The piston is moving down as air is being sucked in by the downward motion against the piston.

- Compression: This stroke begins at B.D.C, or just at the end of the suction stroke, and ends at T.D.C. In this stroke the piston compresses the air-fuel mixture in preparation for ignition during the power stroke (below). Both the intake and exhaust valves are closed during this stage.

- Combustion: Also known as power or ignition. This is the start of the second revolution of the four stroke cycle. At this point the crankshaft has completed a full 360 degree revolution. While the piston is at T.D.C. (the end of the compression stroke) the compressed air-fuel mixture is ignited by a spark plug (in a gasoline engine) or by heat generated by high compression (diesel engines), forcefully returning the piston to B.D.C. This stroke produces mechanical work from the engine to turn the crankshaft.

- Exhaust: Also known as outlet. During the *exhaust* stroke, the piston, once again, returns from B.D.C. to T.D.C. while the exhaust valve is open. This action expels the spent air-fuel mixture through the exhaust valve.

These four strokes can be remembered by the colloquial phrase, "Suck, Squeeze, Bang, Blow".

Thermodynamic Analysis

The thermodynamic analysis of the actual four-stroke and two-stroke cycles is not a simple task. However, the analysis can be simplified significantly if air standard assumptions are utilized. The resulting cycle, which closely resembles the actual operating conditions, is the Otto cycle.

During normal operation of the engine, as the air/fuel mixture is being compressed, an electric spark is created to ignite the mixture. At low rpm this occurs close to TDC (Top Dead Centre). As engine rpm rises, the speed of the flame front does not change so the spark point is advanced earlier in the cycle to allow a greater proportion of the cycle for the charge to combust before the power stroke commences. This advantage is reflected in the various Otto engine designs; the atmospheric (non-compression) engine operates at 12% efficiency whereas the compressed-charge engine has an operating efficiency around 30%.

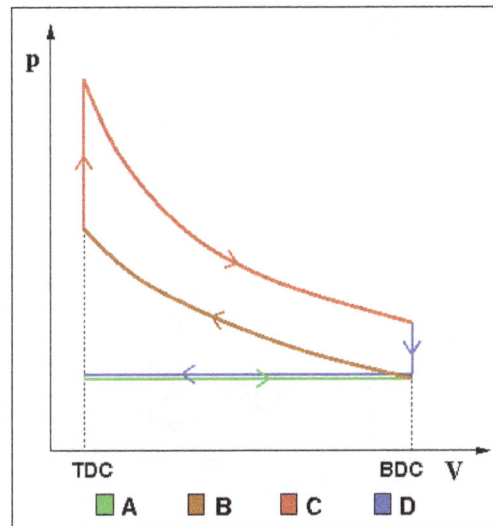

The idealized four-stroke Otto cycle p-V diagram: the intake (A) stroke is performed by an iso-baric expansion, followed by the compression (B) stroke, performed by an adiabatic compression. Through the combustion of fuel an isochoric process is produced, followed by an adiabatic expansion, characterizing the power (C) stroke. The cycle is closed by an isochoric process and an isobaric compression, characterizing the exhaust (D) stroke.

Fuel Considerations

A problem with compressed charge engines is that the temperature rise of the compressed charge can cause pre-ignition. If this occurs at the wrong time and is too energetic, it can damage the engine. Different fractions of petroleum have widely varying flash points (the temperatures at which the fuel may self-ignite). This must be taken into account in engine and fuel design.

The tendency for the compressed fuel mixture to ignite early is limited by the chemical composition of the fuel. There are several grades of fuel to accommodate differing performance levels of engines. The fuel is altered to change its self ignition temperature. There are several ways to do this. As engines are designed with higher compression ratios the result is that pre-ignition is much more likely to occur since the fuel mixture is compressed to a higher temperature prior to deliberate ignition. The higher temperature more effectively evaporates fuels such as gasoline, which increases the efficiency of the compression engine. Higher Compression ratios also means that the distance that the piston can push to produce power is greater (which is called the Expansion ratio).

The octane rating of a given fuel is a measure of the fuel's resistance to self-ignition. A fuel with a higher numerical octane rating allows for a higher compression ratio, which extracts more energy from the fuel and more effectively converts that energy into useful work while at the same time preventing engine damage from pre-ignition. High Octane fuel is also more expensive.

Diesel engines by their nature do not have concerns with pre-ignition. They have a concern with whether or not combustion can be started. The description of how likely Diesel fuel is to ignite is called the Cetane rating. Because Diesel fuels are of low volatility, they can be very hard to start when cold. Various techniques are used to start a cold Diesel engine, the most common being the use of a glow plug.

Design and Engineering Principles

Power Output Limitations

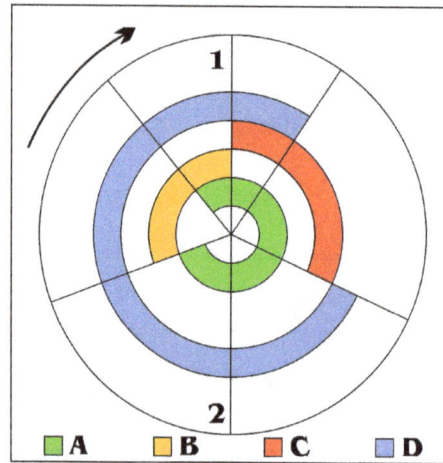

In figure, the four-stroke cycle.

1. TDC

2. BDC

    A: Intake

    B: Compression

    C: Power

    D: Exhaust

The maximum amount of power generated by an engine is determined by the maximum amount of air ingested. The amount of power generated by a piston engine is related to its size (cylinder volume), whether it is a two-stroke engine or four-stroke design, volumetric efficiency, losses, air-to-fuel ratio, the calorific value of the fuel, oxygen content of the air and speed (RPM). The speed is ultimately limited by material strength and lubrication. Valves, pistons and connecting rods suffer severe acceleration forces. At high engine speed, physical breakage and piston ring flutter can occur, resulting in power loss or even engine destruction. Piston ring flutter occurs when the rings oscillate vertically within the piston grooves they reside in. Ring flutter compromises the seal between the ring and the cylinder wall, which causes a loss of cylinder pressure and power. If an engine spins too quickly, valve springs cannot act quickly enough to close the valves. This is commonly referred to as 'valve float', and it can result in piston to valve contact, severely damaging the engine. At high speeds the lubrication of piston cylinder wall interface tends to break down. This limits the piston speed for industrial engines to about 10 m/s.

Intake/Exhaust Port Flow

The output power of an engine is dependent on the ability of intake (air–fuel mixture) and exhaust matter to move quickly through valve ports, typically located in the cylinder head. To increase an engine's output power, irregularities in the intake and exhaust paths, such as casting flaws, can be removed and with the aid of an air flow bench, the radii of valve port turns and valve seat configuration can be modified to reduce resistance. This process is called porting, and it can be done by hand or with a CNC machine.

Waste Heat Recovery of an Internal Combustion Engine

An internal combustion engine is on average capable of converting only 40-45% of supplied energy into mechanical work. A large part of the waste energy is in the form of heat that is released to the environment through coolant, fins etc. If we could somehow recover the waste heat we can improve the engine's performance. It has been found that even if 6% of the entirely wasted heat is recovered it can increase the engine efficiency greatly.

Many methods have been devised in order to extract waste heat out of an engine exhaust and use it further to extract some useful work, decreasing the exhaust pollutants at the same time. Use of the Rankine Cycle, turbocharging and thermoelectric generation can be very useful as a waste heat recovery system.

Though these systems are used more frequently some issues, like their low efficiency at lower heat supply rates and high pumping losses, remain a cause of concern.

Supercharging

One way to increase engine power is to force more air into the cylinder so that more power can be produced from each power stroke. This can be done using some type of air compression device known as a supercharger, which can be powered by the engine crankshaft.

Supercharging increases the power output limits of an internal combustion engine relative to its displacement. Most commonly, the supercharger is always running, but there have been designs that allow it to be cut out or run at varying speeds (relative to engine speed). Mechanically driven supercharging has the disadvantage that some of the output power is used to drive the supercharger, while power is wasted in the high pressure exhaust, as the air has been compressed twice and then gains more potential volume in the combustion but it is only expanded in one stage.

Turbocharging

A turbocharger is a supercharger that is driven by the engine's exhaust gases, by means of a turbine. A turbocharger is incorporated into the exhaust system of a vehicle to make use of the expelled exhaust. It consists of a two piece, high-speed turbine assembly with one side that compresses the intake air, and the other side that is powered by the exhaust gas outflow.

When idling, and at low-to-moderate speeds, the turbine produces little power from the small exhaust volume, the turbocharger has little effect and the engine operates nearly in a naturally aspirated manner. When much more power output is required, the engine speed and throttle opening are increased until the exhaust gases are sufficient to 'spool up' the turbocharger's turbine to start compressing much more air than normal into the intake manifold. Thus, additional power (and speed) is expelled through the function of this turbine.

Turbocharging allows for more efficient engine operation because it is driven by exhaust pressure that would otherwise be (mostly) wasted, but there is a design limitation known as turbo lag. The increased engine power is not immediately available due to the need to sharply increase engine RPM, to build up pressure and to spin up the turbo, before the turbo starts to do any useful air compression. The increased intake volume causes increased exhaust and spins the turbo faster, and so forth until steady high power operation is reached. Another difficulty is that the higher exhaust pressure causes the exhaust gas to transfer more of its heat to the mechanical parts of the engine.

Rod and Piston-to-stroke Ratio

The rod-to-stroke ratio is the ratio of the length of the connecting rod to the length of the piston stroke. A longer rod reduces sidewise pressure of the piston on the cylinder wall and the stress forces, increasing engine life. It also increases the cost and engine height and weight.

A "square engine" is an engine with a bore diameter equal to its stroke length. An engine where the bore diameter is larger than its stroke length is an oversquare engine; conversely, an engine with a bore diameter that is smaller than its stroke length is an undersquare engine.

Valve Train

The valves are typically operated by a camshaft rotating at half the speed of the crankshaft. It has a series of cams along its length, each designed to open a valve during the appropriate part of an intake or exhaust stroke. A tappet between valve and cam is a contact surface on which the cam slides to open the valve. Many engines use one or more camshafts "above" a row (or each row) of cylinders, in which each cam directly actuates a valve through a flat tappet. In other engine designs the camshaft is in the crankcase, in which case each cam usually contacts a push rod, which contacts a rocker arm that opens a valve, or in case of a flathead engine a push rod is not necessary. The overhead cam design typically allows higher engine speeds because it provides the most direct path between cam and valve.

Valve Clearance

Valve clearance refers to the small gap between a valve lifter and a valve stem that ensures that the valve completely closes. On engines with mechanical valve adjustment, excessive clearance causes noise from the valve train. A too small valve clearance can result in the valves not closing properly, this results in a loss of performance and possibly overheating of exhaust valves. Typically, the clearance must be readjusted each 20,000 miles (32,000 km) with a feeler gauge.

Most modern production engines use hydraulic lifters to automatically compensate for valve train component wear. Dirty engine oil may cause lifter failure.

Energy Balance

Otto engines are about 30% efficient; in other words, 30% of the energy generated by combustion is converted into useful rotational energy at the output shaft of the engine, while the remainder being losses due to waste heat, friction and engine accessories. There are a number of ways to recover some of the energy lost to waste heat. The use of a Turbocharger in Diesel engines is very effective by boosting incoming air pressure and in effect, provides the same increase in performance as having more displacement. The Mack Truck company, decades ago, developed a turbine system that converted waste heat into kinetic energy that it fed back into the engine's transmission. In 2005, BMW announced the development of the turbosteamer, a two-stage heat-recovery system similar to the Mack system that recovers 80% of the energy in the exhaust gas and raises the efficiency of an Otto engine by 15%. By contrast, a six-stroke engine may reduce fuel consumption by as much as 40%.

Modern engines are often intentionally built to be slightly less efficient than they could otherwise be. This is necessary for emission controls such as exhaust gas recirculation and catalytic converters that reduce smog and other atmospheric pollutants. Reductions in efficiency may be counteracted with an engine control unit using lean burn techniques.

In the United States, the Corporate Average Fuel Economy mandates that vehicles must achieve an average of 34.9 mpg_{-US} (6.7 L/100 km; 41.9 mpg_{-imp}) compared to the current standard of 25 mpg_{-US} (9.4 L/100 km; 30.0 mpg_{-imp}). As automakers look to meet these standards by 2016, new ways of engineering the traditional internal combustion engine (ICE) have to be considered. Some potential solutions to increase fuel efficiency to meet new mandates include firing after the piston is farthest from the crankshaft, known as top dead centre, and applying the Miller cycle. Together, this redesign could significantly reduce fuel consumption and NO_x emissions.

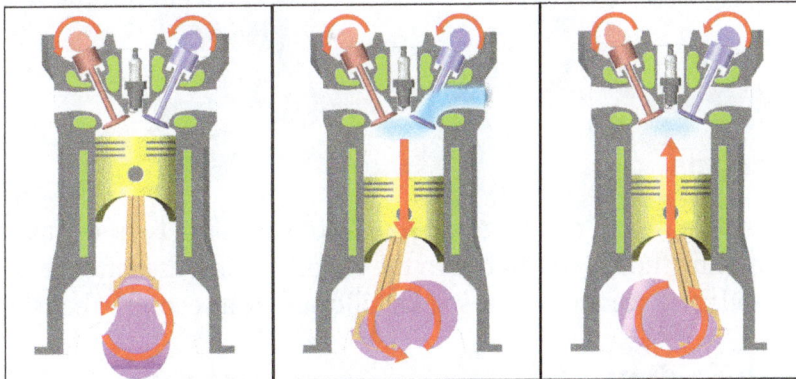

Starting position, intake stroke, and compression stroke.

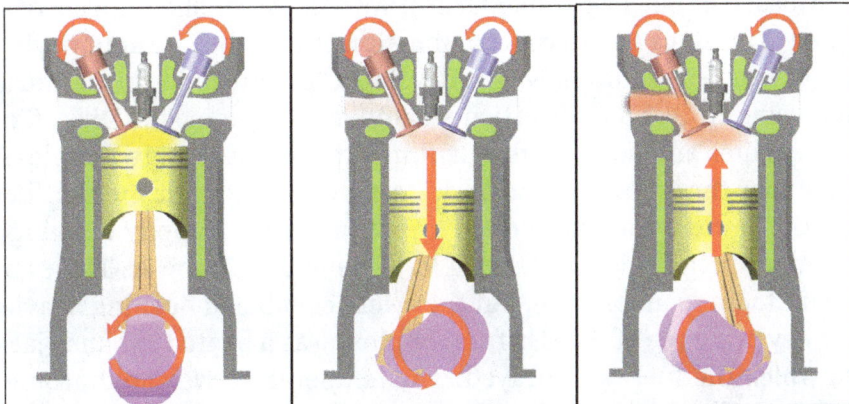

Ignition of fuel, power stroke, and exhaust stroke.

Six-stroke Engine

The term six-stroke engine has been applied to a number of alternative internal combustion engine designs that attempt to improve on traditional two-stroke and four-stroke engines. Claimed advantages may include increased fuel efficiency, reduced mechanical complexity and reduced emissions. These engines can be divided into two groups based on the number of pistons that contribute to the six strokes.

In the single-piston designs, the engine captures the heat lost from the four-stroke Otto cycle or Diesel cycle and uses it to drive an additional power and exhaust stroke of the piston in the same cylinder in an attempt to improve fuel-efficiency and assist with engine cooling. The pistons in this type of six-stroke engine go up and down three times for each injection of fuel. These designs use either steam or air as the working fluid for the additional power stroke.

The designs in which the six strokes are determined by the interactions between two pistons are more diverse. The pistons may be opposed in a single cylinder or may reside in separate cylinders. Usually one cylinder makes two strokes while the other makes four strokes giving six piston movements per cycle. The second piston may be used to replace the valve mechanism of a conventional engine, which may reduce mechanical complexity and enable an increased compression ratio by eliminating hotspots that would otherwise limit compression. The second piston may also be used to increase the expansion ratio, decoupling it from the compression ratio. Increasing the expansion ratio in this way can increase thermodynamic efficiency in a similar manner to the Miller or Atkinson cycle.

Engine Types

Single-piston Designs

These designs use a single piston per cylinder, like a conventional two- or four-stroke engine. A secondary, non-detonating fluid is injected into the chamber, and the leftover heat from combustion causes it to expand for a second power stroke followed by a second exhaust stroke.

Griffin Six-stroke Engine

In 1883, the Bath-based engineer Samuel Griffin was an established maker of steam and gas engines. He wished to produce an internal combustion engine, but without paying the licensing costs of the Otto patents. His solution was to develop a "patent slide valve" and a single-acting six-stroke engine using it. By 1886, Scottish steam locomotive maker Dick, Kerr & Co. saw a future in large oil engines and licensed the Griffin patents. These were double-acting, tandem engines and sold under the name "Kilmarnock". A major market for the Griffin engine was in electricity generation, where they developed a reputation for happily running light for long periods, then suddenly being able to take up a large demand for power. Their large heavy construction didn't suit them to mobile use, but they were capable of burning heavier and cheaper grades of oil. The key principle of the "Griffin Simplex" was a heated exhaust-jacketed external vapouriser, into which the fuel was sprayed. The temperature was held around 550 °F (288 °C), sufficient to physically vapourise the oil but not to break it down chemically. This fractional distillation supported the use of heavy oil fuels, the unusable tars and asphalts separating out in the vapouriser. Hot-bulb ignition was used, which Griffin termed the "catathermic igniter", a small isolated cavity connected to the combustion chamber. The spray injector had an adjustable inner nozzle for the air supply, surrounded by an annular casing for the oil, both oil and air entering at 20 psi (140 kPa) pressure, and being regulated by a governor. Griffin went out of business in 1923. Only two known examples of a Griffin six-stroke engine survive. One is in the Anson Engine Museum. The other was built in 1885 and for some years was in the Birmingham Museum of Science and Technology, but in 2007 it returned to Bath and the Museum of Bath at Work.

The Kerr engine at the Anson Engine Museum.

Dyer Six-stroke Engine

Leonard Dyer invented a six-stroke internal combustion water-injection engine in 1915, very similar to Crower's design. A dozen more similar patents have been issued since.

Dyer's six-stroke engine features:

- No cooling system required.
- Improves a typical engine's fuel consumption.
- Requires a supply of pure water to act as the medium for the second power stroke.
- Extracts the additional power from the expansion of steam.

Bajulaz Six-stroke Engine

The Bajulaz six-stroke engine is similar to a regular combustion engine in design. There are, however, modifications to the cylinder head, with two supplementary fixed-capacity chambers: a combustion chamber and an air-preheating chamber above each cylinder. The combustion chamber receives a charge of heated air from the cylinder; the injection of fuel begins an isochoric (constant-volume) burn, which increases the thermal efficiency compared to a burn in the cylinder. The high pressure achieved is then released into the cylinder to work the power or expansion stroke. Meanwhile, a second chamber, which blankets the combustion chamber, has its air content heated to a high degree by heat passing through the cylinder wall. This heated and pressurized air is then used to power an additional stroke of the piston.

The claimed advantages of the engine include reduction in fuel consumption by at least 40%, two expansion strokes in six strokes, multi-fuel usage capability, and a dramatic reduction in pollution.

The Bajulaz six-stroke engine features claimed are:

- Reduction in fuel consumption by at least 40%.

- Two expansion (work) strokes in six strokes.

- Multifuel, including liquefied petroleum gas.

- Dramatic reduction in air pollution.

- Costs comparable to those of a four-stroke engine.

Velozeta Six-stroke Engine

In a Velozeta engine, fresh air is injected into the cylinder during the exhaust stroke, which expands by heat and therefore forces the piston down for an additional stroke. The valve overlaps have been removed, and the two additional strokes using air injection provide for better gas scavenging. The engine seems to show 40% reduction in fuel consumption and dramatic reduction in air pollution. Its Power-to-weight ratio is slightly less than that of a four-stroke gasoline engine. The engine can run on a variety of fuels, ranging from gasoline and diesel fuel to LPG. An altered engine shows a 65% reduction in carbon monoxide pollution when compared with the four-stroke engine from which it was developed. The engine was developed in 2005 by a team of mechanical engineering students, Mr. U Krishnaraj, Mr. Boby Sebastian, Mr. Arun Nair and Mr. Aaron Joseph George of the College of Engineering, Trivandrum.

Crower Six-stroke Engine

In a six-stroke engine prototyped in the United States by Bruce Crower, water is injected into the cylinder after the exhaust stroke and is instantly turned to steam, which expands and forces the piston down for an additional power stroke. Thus, waste heat that requires an air or water cooling system to discharge in most engines is captured and put to use driving the piston. Crower estimated that his design would reduce fuel consumption by 40% by generating the same power output at a lower rotational speed. The weight associated with a cooling system could be eliminated, but that would be balanced by a need for a water tank in addition to the normal fuel tank.

The Crower six-stroke engine was an experimental design that attracted media attention in 2006 because of an interview given by the 75-year-old American inventor, who has applied for a patent on his design. That patent application was subsequently abandoned.

Opposed-piston Designs

These designs use two pistons per cylinder operating at different rates, with combustion occurring between the pistons.

Beare Head

This design was developed by Malcolm Beare of Australia. The technology combines a four-stroke engine bottom end with an opposed piston in the cylinder head working at half the cyclical rate of the bottom piston. Functionally, the second piston replaces the valve mechanism of a conventional engine. Claimed benefits include a 9% increase in power, and improved thermodynamic efficiency through an increased compression ratio enabled by the elimination of the hot exhaust valve.

M4+2

The idea was developed at the Silesian University of Technology, Poland, under the leadership of dr inż. Adam Ciesiołkiewicz. It was granted patent nr 195052 by the Polish Patent Office.

The M4+2 engines have much in common with the Beare-head engines, combining two opposed pistons in the same cylinder. One piston works at half the cyclical rate of the other, but while the main function of the second piston in a Beare-head engine is to replace the valve mechanism of a conventional four-stroke engine, the M4+2 takes the principle one step further. The double-piston combustion engine's work is based on the cooperation of both modules. The air load change takes place in the two-stroke section of the engine. The piston of the four-stroke section is an air load exchange aiding system, working as a system of valves. The cylinder is filled with air or with an air-fuel mixture. The filling process takes place at overpressure by the slide inlet system. The exhaust gases are removed as in the classical two-stroke engine, by exhaust windows in the cylinder. The fuel is supplied into the cylinder by a fuel-injection system. Ignition is realized by two spark plugs. The effective power output of the double-piston engine is transferred by two crankshafts. The characteristic feature of this engine is an opportunity of continuous change of cylinder capacity and compression rate during engine work by changing the piston's location. The mechanical and thermodynamical models were meant for double-piston engines, which enable to draw up new theoretical thermodynamic cycle for internal combustion double-pistons engine.

Other Two-piston Designs

Piston-charger Engine

In this engine, similar in design to the Beare head, a "piston charger" replaces the valve system. The piston charger charges the main cylinder and simultaneously regulates the inlet and the outlet aperture, leading to no loss of air and fuel in the exhaust. In the main cylinder, combustion takes place every turn as in a two-stroke engine and lubrication as in a four-stroke. Fuel injection can take place in the piston charger, in the gas-transfer channel or in the combustion chamber. It is also possible to charge two working cylinders with one piston charger. The combination of compact design for the combustion chamber together with no loss of air and fuel is claimed to give the engine more torque, more power and better fuel consumption. The benefit of fewer moving parts and design is claimed to lead to lower manufacturing costs. Good for hybrid technology and stationary engines. The engine is claimed to be suited to alternative, fuels since there is no corrosion or deposits left on valves. The six strokes are:

- Aspiration

- Precompression

- Gas transfer

- Compression

- Ignition

- Ejection

Ilmor/Schmitz Five-stroke

This design was invented by Belgian engineer Gerhard Schmitz, and has been prototyped by Ilmor Engineering.

These designs use two (or 4, 6, 8) cylinders with a conventional Otto four-stroke cycle. An additional piston (in its own cylinder) is shared by the two Otto cycle cylinders. The exhaust from the Otto cycle cylinder is directed into the shared cylinder, where it is expanded generating additional work. This is in some respects similar to the operation of a compound steam engine, with the Otto cycle cylinders being the high-pressure stage and the shared cylinder the low pressure stage. The operation of the engine is thus:

HP1 (Otto)	LP (shared)	HP2 (Otto)
exhaust	expansion (power)	compression
intake	exhaust	power
compression	expansion (power)	exhaust
power	exhaust	intake

The designers consider this to be a five-stroke design, regarding the simultaneous HP exhaust stroke and LP expansion stroke as a single stroke. This design provides higher fuel efficiency due to the higher overall expansion ratio of the combined cylinders. Expansion ratios comparable to diesel engines can be achieved, while still using gasoline (petrol) fuel. Five-stroke engines allegedly are lighter and have higher power density than diesel engines.

Revetec Engines

The controlled combustion engines, designed by Bradley Howell-Smith of Australian firm Revetec Holdings Pty Ltd, use opposed pairs of pistons to drive a pair of counter-rotating three-lobed cams through bearings. These elements replace the conventional crankshaft and connecting-rods, which enables the motion of the pistons to be purely axial, so that most of the power otherwise wasted on lateral motion of the con-rods is effectively transferred to the output shaft. This gives six power strokes per revolution of the shaft (spread across a pair of pistons). An independent test measured the BSFC of Revetec's X4v2 prototype gasoline engine at 212g/kW-h (corresponding to an energy efficiency of 38.6%). Any even number of pistons can be used, in boxer or X configurations; the three lobes of the cams can be replaced by any other odd number greater than one; and the geometry of the cams can be changed to suit the needs of the target fuels and applications of the engines. Such variants may have ten or more strokes per cycle.

Atkinson Cycle

The Atkinson-cycle engine is a type of internal combustion engine invented by James Atkinson in 1882. The Atkinson cycle is designed to provide efficiency at the expense of power density.

A modern variation of this approach is used in some modern automobile engines. While originally seen exclusively in hybrid electric applications such as the earlier-generation Toyota Prius, later hybrids and some non-hybrid vehicles now feature engines with variable valve timing, which can run in the Atkinson cycle as a part-time operating regimen, giving good economy while running in Atkinson cycle, and conventional power density when running as a conventional, Otto cycle engine.

Design

Atkinson produced three different designs that had a short compression stroke and a longer expansion stroke. The first Atkinson-cycle engine, the *differential engine*, used opposed pistons. The second and most well-known design, was the *cycle engine*, which used an over-center arm to create four piston strokes in one crankshaft revolution. The reciprocating engine had the intake, compression, power, and exhaust strokes of the four-stroke cycle in a single turn of the crankshaft, and was designed to avoid infringing certain patents covering Otto-cycle engines. Atkinson's third and final engine, the *utilite engine*, operated much like any two-stroke engine.

The common thread throughout Atkinson's designs is that the engines have an expansion stroke that is longer than the compression stroke, and by this method the engine achieves greater thermal efficiency than a traditional piston engine. Atkinson's engines were produced by the British Gas Engine Company and also licensed to other overseas manufacturers.

Many modern engines now use unconventional valve timing to produce the effect of a shorter compression stroke/longer power stroke. Miller applied this technique to the four-stroke engine, so it is sometimes referred as the Atkinson/Miller cycle, US patent 2817322 dated Dec 24, 1957. In 1888, Charon filed a French patent and displayed an engine at the Paris Exhibition in 1889. The Charon gas engine (four-stroke) used a similar cycle to Miller, but without a supercharger. It is referred to as the "Charon cycle".

Modern engine designers are realizing the potential fuel-efficiency improvements the Atkinson-type cycle can provide.

Atkinson Differential Engine

The first implementation of the Atkinson cycle was in 1882; unlike later versions, it was arranged as an opposed piston engine, the Atkinson differential engine. In this, a single crankshaft was connected to two opposed pistons through a toggle-jointed linkage that had a nonlinearity; for half a revolution, one piston remained almost stationary while the other approached it and returned, and then for the next half revolution, the second-mentioned piston was almost stationary while the first approached and returned.

Thus, in each revolution, one piston provided a compression stroke and a power stroke, and then the other piston provided an exhaust stroke and a charging stroke. As the power piston remained withdrawn during exhaust and charging, it was practical to provide exhaust and charging using valves behind a port that was covered during the compression stroke and the power stroke, and so the valves did not need to resist high pressure and could be of the simpler sort used in many steam engines, or even reed valves.

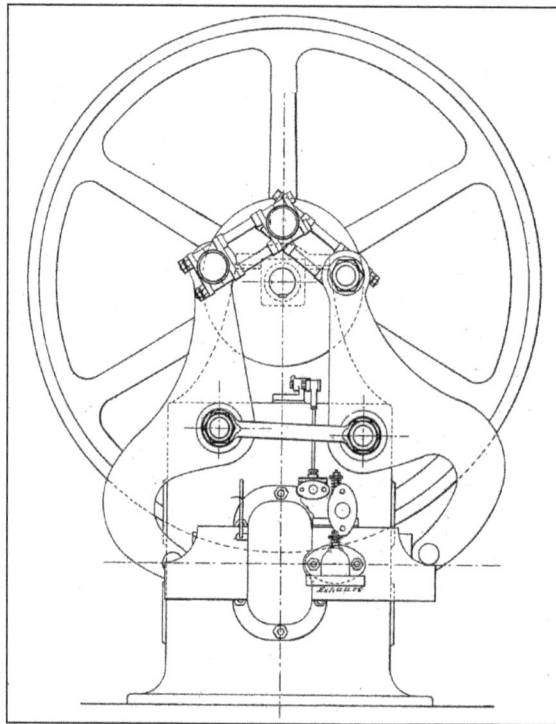

Patent drawing of the Atkinson "Differential Engine".

Atkinson Cycle Engine

The next engine designed by Atkinson in 1887 was named the "Cycle Engine" This engine used poppet valves, a cam, and an over-center arm to produce four piston strokes for every revolution of the crankshaft. The intake and compression strokes were significantly shorter than the expansion and exhaust strokes.

The "Cycle" engines were produced and sold for several years by the British Engine Company. Atkinson also licensed production to other manufacturers. Sizes ranged from a few up to 100 horsepower.

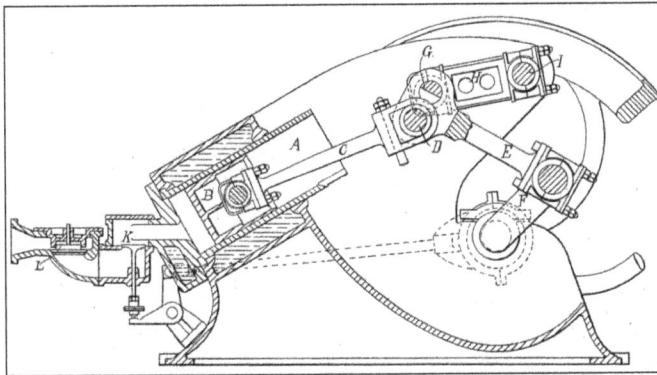

Atkinson gas engine as shown in US Patent 367496.

Atkinson Utilite Engine

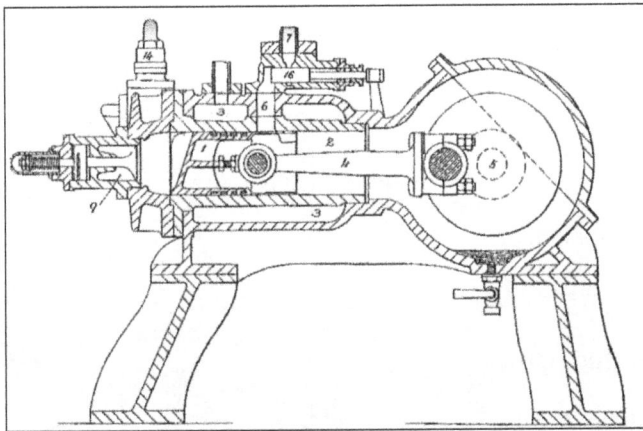

Atkinson's Utilite engine.

Atkinson's third design was named the "Utilite Engine". Atkinson's "Cycle" engine was efficient; however, its linkage was difficult to balance for high speed operation. Atkinson realized an improvement was needed to make his cycle more applicable as a higher-speed engine.

With this new design, Atkinson was able to eliminate the linkages and make a more conventional, well balanced engine capable of operating at speeds up to 600 rpm and capable of producing power every revolution yet he preserved all of the efficiency of his "Cycle Engine" having a proportionally short compression stroke and a longer expansion stroke. The Utilite operates much like a standard two-stroke except that the exhaust port is located at about the middle of the stroke.

During the expansion/power stroke, a cam-operated valve (which remains closed until the piston nears the end of the stroke) prevents pressure from escaping as the piston moves past the exhaust port. The exhaust valve is opened near the bottom of the stroke; it remains open as the piston heads back toward compression, letting fresh air charge the cylinder and exhaust escape until the port is covered by the piston.

After the exhaust port is covered the piston begins to compress the remaining air in the cylinder. A small piston fuel pump injects liquid during compression. The ignition source was likely a hot tube as in Atkinson's other engines. This design resulted in a two-stroke engine with a short compression and longer expansion stroke.

The Utilite Engine tested as even more efficient than Atkinson's previous "differential" and "cycle" designs. Very few were produced, and none are known to survive. The British patent is from 1892, #2492. No US patent for the Utilite Engine is known.

Ideal Thermodynamic Cycle

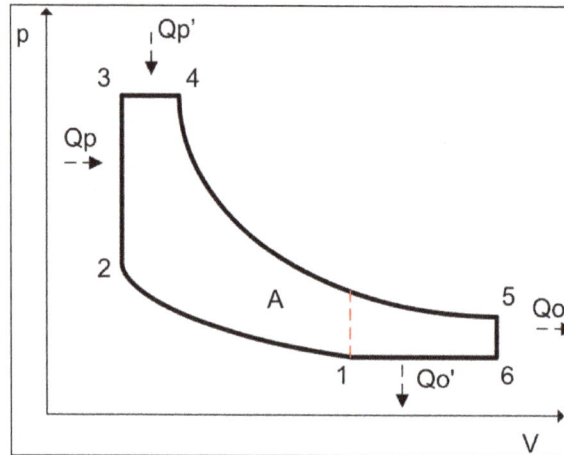

Atkinson gas cycle.

The ideal Atkinson cycle consists of:

- 1–2 Isentropic, or reversible, adiabatic compression.
- 2–3 Isochoric heating (Qp).
- 3–4 Isobaric heating (Qp').
- 4–5 Isentropic expansion.
- 5–6 Isochoric cooling (Qo).
- 6–1 Isobaric cooling (Qo').

Modern Atkinson-cycle Engines

A small engine with Atkinson-style linkages between the piston and flywheel.
Modern Atkinson-cycle engines do away with this complex energy path.

In the late 20th century, the term "Atkinson cycle" began to be used to describe a modified Otto-cycle engine—in which the intake valve is held open longer than normal, allowing a reverse flow of intake air into the intake manifold. This "simulated" Atkinson cycle is most notably used in the Toyota 1NZ-FXE engine from the early Prius.

The effective compression ratio is reduced—for the time the air is escaping the cylinder freely rather than being compressed—but the *expansion* ratio is unchanged (i.e., the compression ratio is smaller than the expansion ratio). The goal of the modern Atkinson cycle is to make the pressure in the combustion chamber at the end of the power stroke equal to atmospheric pressure. When this occurs, all available energy has been obtained from the combustion process. For any given portion of air, the greater expansion ratio converts more energy from heat to useful mechanical energy— meaning the engine is more efficient.

The disadvantage of the four-stroke Atkinson-cycle engine versus the more common Otto-cycle engine is reduced power density. Due to a smaller portion of the compression stroke being devoted to compressing the intake air, an Atkinson-cycle engine does not take in as much air as would a similarly designed and sized Otto-cycle engine. Four-stroke engines of this type that use the same type of intake valve motion but using forced induction to make up for the loss of power density are known as Miller-cycle engines.

Rotary Atkinson-cycle Engine

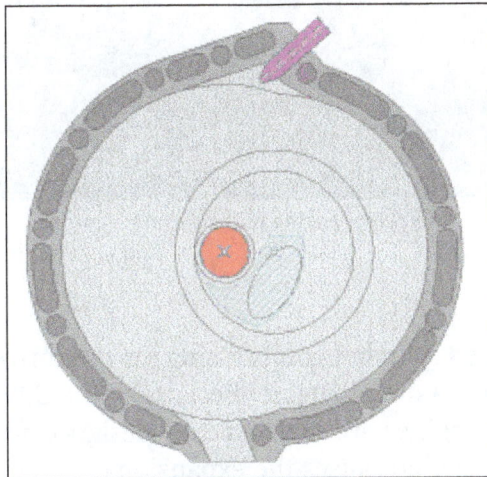

Rotary Atkinson-cycle engine.

The Atkinson-cycle can be used in a rotary engine. In this configuration, an increase in both power and efficiency can be achieved when compared to the Otto cycle. This type of engine retains the one power phase per revolution, together with the different compression and expansion volumes of the original Atkinson-cycle.

Exhaust gases are expelled from the engine by compressed-air scavenging. This modification of the Atkinson-cycle allows the use of alternative fuels such as diesel and hydrogen.

Disadvantages of this design include the requirement that rotor tips seal very tightly on the outer housing wall and the mechanical losses suffered through friction between rapidly oscillating parts of irregular shape.

Petrol Engine

A petrol engine (known as a gasoline engine is an internal combustion engine with spark-ignition, designed to run on petrol (gasoline) and similar volatile fuels.

In most petrol engines, the fuel and air are usually mixed after compression (although some modern petrol engines now use cylinder-direct petrol injection). The pre-mixing was formerly done in a carburetor, but now it is done by electronically controlled fuel injection, except in small engines where the cost/complication of electronics does not justify the added engine efficiency. The process differs from a diesel engine in the method of mixing the fuel and air, and in using spark plugs to initiate the combustion process. In a diesel engine, only air is compressed (and therefore heated), and the fuel is injected into very hot air at the end of the compression stroke, and self-ignites.

W16 petrol engine of the Bugatti Veyron.

Compression Ratio

With both air and fuel in a closed cylinder, compressing the mixture too much poses the danger of auto-ignition — or behaving like a diesel engine. Because of the difference in burn rates between the two different fuels, petrol engines are mechanically designed with different timing than diesels, so to auto-ignite a petrol engine causes the expansion of gas inside the cylinder to reach its greatest point before the cylinder has reached the "top dead center" (T.D.C) position. Spark plugs are typically set statically or at idle at a minimum of 10 degrees or so of crankshaft rotation before the piston reaches T.D.C, but at much higher values at higher engine speeds to allow time for the fuel-air charge to substantially complete combustion before too much expansion has occurred - gas expansion occurring with the piston moving down in the power stroke. Higher octane petrol burns slower, therefore it has a lower propensity to auto-ignite and its rate of expansion is lower. Thus, engines designed to run high-octane fuel exclusively can achieve higher compression ratios.

Most modern automobile petrol engines generally have a compression ratio of 10.0:1 to 13.5:1. Engines with a knock sensor can and usually have C.R higher than 11.1:1 and approaches 14.0:1 (for high octane fuel and usually with direct fuel injection) and engines without a knock sensor generally have C.R of 8.0:1 to 10.5:1.

Speed and Efficiency

Petrol engines run at higher rotation speeds than diesels, partially due to their lighter pistons, connecting rods and crankshaft (a design efficiency made possible by lower compression ratios) and due to petrol burning more quickly than diesel.

Because pistons in petrol engines tend to have much shorter strokes than pistons in diesel engines, typically it takes less time for a piston in a petrol engine to complete its stroke than a piston in a diesel engine. However, the lower compression ratios of petrol engines give petrol engines lower efficiency than diesel engines.

Typically, most petrol engines have approximately 20%(avg.) thermal efficiency, which is nearly half of diesel engines. However some newer engines are reported to be much more efficient(thermal efficiency up to 38%) than previous spark-ignition engines.

Applications

Current

Petrol engines have many applications, including:

- Automobiles,

- Motorcycles,

- Aircraft,

- Motorboats,

- Small engines, such as lawn mowers, chainsaws and portable engine-generators.

Design

Cylinder Arrangement

Common cylinder arrangements are from 1 to 6 cylinders in-line or from 2 to 16 cylinders in V-formation. Flat engines – like a V design flattened out – are common in small airplanes and motorcycles and were a hallmark of Volkswagen automobiles into the 1990s. Flat 6s are still used in many modern Porsches, as well as Subarus. Many flat engines are air-cooled. Less common, but notable in vehicles designed for high speeds is the W formation, similar to having 2 V engines side by side. Alternatives include rotary and radial engines the latter typically have 7 or 9 cylinders in a single ring, or 10 or 14 cylinders in two rings.

Cooling

Petrol engines may be air-cooled, with fins (to increase the surface area on the cylinders and cylinder head); or liquid-cooled, by a water jacket and radiator. The coolant was formerly water, but is now usually a mixture of water and either ethylene glycol or propylene glycol. These mixtures have lower freezing points and higher boiling points than pure water and also prevent corrosion, with modern antifreezes also containing lubricants and other additives to protect water pump seals and bearings. The cooling system is usually slightly pressurized to further raise the boiling point of the coolant.

Ignition

Petrol engines use spark ignition and high voltage current for the spark may be provided by a magneto or an ignition coil. In modern car engines the ignition timing is managed by an electronic Engine Control Unit.

Power Measurement

The most common way of engine rating is what is known as the brake power, measured at the flywheel, and given in metric horsepower or kilowatts (metric), or in horsepower (Imperial/USA). This is the actual mechanical power output of the engine in a usable and complete form. The term "brake" comes from the use of a brake in a dynamometer test to load the engine. For accuracy, it is important to understand what is meant by usable and complete. For example, for a car engine, apart from friction and thermodynamic losses inside the engine, power is absorbed by the water pump, alternator, and radiator fan, thus reducing the power available at the flywheel to move the car along. Power is also absorbed by the power steering pump and air conditioner (if fitted), but these are not installed for a power output test or calculation. Power output varies slightly according to the energy value of the fuel, the ambient air temperature and humidity, and the altitude. Therefore, there are agreed standards in the USA and Europe on the fuel to use when testing, and engines are rated at 25 °C (Europe), and 64 °F (USA) at sea level, 50% humidity. Marine engines, as supplied, usually have no radiator fan, and often no alternator. In such cases the quoted power rating does not allow for losses in the radiator fan and alternator. The SAE in USA, and the ISO in Europe, publish standards on exact procedures, and how to apply corrections for non-standard conditions such as altitude above sea level.

Car testers are most familiar with the chassis dynamometer or "rolling road" installed in many workshops. This measures drive wheel brake horsepower, which is generally 15-20% less than the brake horsepower measured at the crankshaft or flywheel on an engine dynamometer.

Advantages of Reciprocating Internal Combustion Engines

- Unlike steam turbine no heat exchanger required in reciprocating internal combustion engines, which results in mechanical simplicity and improved power plant efficiency.

- As compared to steam or gas turbines, the average working temperatures of reciprocating internal combustion engines is very low. The reason is that the peak temperature exists for a very less time (in a cycle). This results in the application of high temperatures resulting in high thermal efficiency of reciprocating internal combustion engines.

- In reciprocating internal combustion engines high thermal efficiency can be obtained by the application of moderate maximum working pressure of the fluid. This results in less weight to power ratio in reciprocating internal combustion engines. Hence for the same power reciprocating internal combustion engines have less weight compared a steam power plant.

- It is also possible with reciprocating internal combustion engines to generate very small power output (even a fraction of kilowatt) with reasonable thermal efficiency and cost.

Disadvantages of Reciprocating Internal Combustion Engines

- Due to reciprocating parts more vibrations are produced in reciprocating internal combustion engines. This results in decreased lifespan.

- It is also not possible to use variety of fuels with reciprocating internal combustion engines (like coal, wood etc).

- Fuels used in reciprocating internal combustion engines are relatively more expensive than fuels used in steam or gas power plant.

- Due to reciprocating parts consumption of lubricants is also high in reciprocating internal combustion engines as compared to steam power plant.

- Reciprocating internal combustion engines are only suitable, when we need low power output. For high power output generally steam power plants are used (like generation of electricity etc).

Diesel Engine

The diesel engine (also known as a compression-ignition or CI engine), named after Rudolf Diesel, is an internal combustion engine in which ignition of the fuel is caused by the elevated temperature of the air in the cylinder due to the mechanical compression (adiabatic compression). This contrasts with spark-ignition engines such as a petrol engine (gasoline engine) or gas engine (using a gaseous fuel as opposed to petrol), which use a spark plug to ignite an air-fuel mixture.

Diesel engine built by Langen & Wolf under licence.

Diesel engines work by compressing only the air. This increases the air temperature inside the cylinder to such a high degree that atomised diesel fuel injected into the combustion chamber ignites spontaneously. With the fuel being injected into the air just before combustion, the dispersion of the fuel is uneven; this is called a heterogeneous air-fuel mixture. The torque a diesel engine produces is controlled by manipulating the air ratio; instead of throttling the intake air, the diesel engine relies on altering the amount of fuel that is injected, and the air ratio is usually high.

The diesel engine has the highest thermal efficiency (engine efficiency) of any practical internal or external combustion engine due to its very high expansion ratio and inherent lean burn which enables heat dissipation by the excess air. A small efficiency loss is also avoided compared to two-stroke non-direct-injection gasoline engines since unburned fuel is not present at valve overlap and therefore no fuel goes directly from the intake/injection to the exhaust. Low-speed diesel engines (as used in ships and other applications where overall engine weight is relatively unimportant) can reach effective efficiencies of up to 55%.

Diesel engines may be designed as either two-stroke or four-stroke cycles. They were originally used as a more efficient replacement for stationary steam engines. Since the 1910s they have been used in submarines and ships. Use in locomotives, trucks, heavy equipment and electricity generation plants followed later. In the 1930s, they slowly began to be used in a few automobiles. Since the 1970s, the use of diesel engines in larger on-road and off-road vehicles in the US has increased. According to Konrad Reif, the EU average for diesel cars accounts for half of newly registered cars.

The world's largest diesel engines put in service are 14-cylinder, two-stroke watercraft diesel engines; they produce a peak power of almost 100 MW each.

Operating Principle

Characteristics

The characteristics of a diesel engine are:

- Compression ignition: Due to almost adiabatic compression, the fuel ignites without any ignition-initiating apparatus such as spark plugs.

- Mixture formation inside the combustion chamber: Air and fuel are mixed in the combustion chamber and not in the inlet manifold.

- Engine speed adjustment solely by mixture quality: Instead of throttling the air-fuel mixture, the amount of torque produced (resulting in crankshaft rotational speed differences) is set solely by the mass of injected fuel, always mixed with as much air as possible.

- Heterogeneous air-fuel mixture: The dispersion of air and fuel in the combustion chamber is uneven.

- High air ratio: Due to always running on as much air as possible and not depending on exact mixture of air and fuel, diesel engines have an air-fuel ratio leaner than stochiometric.

- Diffusion flame: At combustion, oxygen first has to diffuse into the flame, rather than having oxygen and fuel already mixed before combustion, which would result in a premixed flame.

- Fuel with high ignition performance: As diesel engines solely rely on compression ignition, fuel with high ignition performance (cetane rating) is ideal for proper engine operation, fuel with a good knocking resistance (octane rating), e.g., petrol, is suboptimal for diesel engines.

Cycle of the Diesel Engine

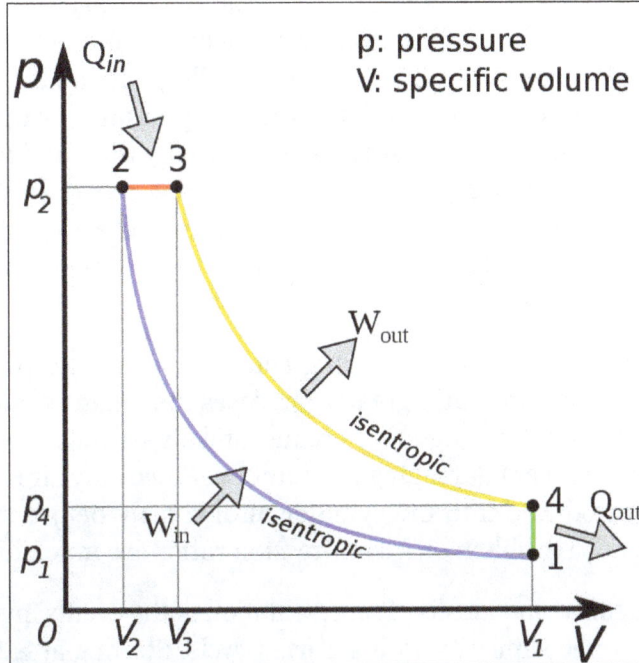

p-V Diagram for the ideal diesel cycle: The cycle follows the numbers 1–4 in clockwise direction. The horizontal axis is volume of the cylinder. In the diesel cycle the combustion occurs at almost constant pressure. On this diagram the work that is generated for each cycle corresponds to the area within the loop.

Diesel engine model, left side.

Diesel engine model, right side.

The diesel internal combustion engine differs from the gasoline powered Otto cycle by using highly compressed hot air to ignite the fuel rather than using a spark plug (*compression ignition* rather than *spark ignition*).

In the diesel engine, only air is initially introduced into the combustion chamber. The air is then compressed with a compression ratio typically between 15:1 and 23:1. This high compression causes the temperature of the air to rise. At about the top of the compression stroke, fuel is injected directly into the compressed air in the combustion chamber. This may be into a (typically toroidal) void in the top of the piston or a *pre-chamber* depending upon the design of the engine. The fuel injector ensures that the fuel is broken down into small droplets, and that the fuel is distributed evenly. The heat of the compressed air vaporises fuel from the surface of the droplets. The vapour is then ignited by the heat from the compressed air in the combustion chamber, the droplets continue to vaporise from their surfaces and burn, getting smaller, until all the fuel in the droplets has been burnt. Combustion occurs at a substantially constant pressure during the initial part of the power stroke. The start of vaporisation causes a delay before ignition and the characteristic diesel knocking sound as the vapour reaches ignition temperature and causes an abrupt increase in pressure above the piston (not shown on the P-V indicator diagram). When combustion is complete the combustion gases expand as the piston descends further; the high pressure in the cylinder drives the piston downward, supplying power to the crankshaft.

As well as the high level of compression allowing combustion to take place without a separate ignition system, a high compression ratio greatly increases the engine's efficiency. Increasing the compression ratio in a spark-ignition engine where fuel and air are mixed before entry to the cylinder is limited by the need to prevent damaging pre-ignition. Since only air is compressed in a diesel engine, and fuel is not introduced into the cylinder until shortly before top dead centre (TDC), premature detonation is not a problem and compression ratios are much higher.

The p–V diagram is a simplified and idealised representation of the events involved in a diesel engine cycle, arranged to illustrate the similarity with a Carnot cycle. Starting at 1, the piston is at bottom dead centre and both valves are closed at the start of the compression stroke; the cylinder contains air at atmospheric pressure. Between 1 and 2 the air is compressed adiabatically – that is without heat transfer to or from the environment – by the rising piston. (This is only approximately true since there will be some heat exchange with the cylinder walls.) During this compression, the volume is reduced, the pressure and temperature both rise. At or slightly before 2 (TDC) fuel is injected and burns in the compressed hot air. Chemical energy is released and this constitutes an injection of thermal energy (heat) into the compressed gas. Combustion and heating occur between 2 and 3. In this interval the pressure remains constant since the piston descends, and the volume increases; the temperature rises as a consequence of the energy of combustion. At 3 fuel injection and combustion are complete, and the cylinder contains gas at a higher temperature than at 2. Between 3 and 4 this hot gas expands, again approximately adiabatically. Work is done on the system to which the engine is connected. During this expansion phase the volume of the gas rises, and its temperature and pressure both fall. At 4 the exhaust valve opens, and the pressure falls abruptly to atmospheric (approximately). This is unresisted expansion and no useful work is done by it. Ideally the adiabatic expansion should continue, extending the line 3–4 to the right until the pressure falls to that of the surrounding air, but the loss of efficiency caused by this unresisted expansion is justified by the practical difficulties involved in recovering it (the engine would have to be much larger). After the opening of the exhaust valve, the exhaust stroke follows, but this (and the following induction stroke) are not shown on the diagram. If shown, they would be represented by a low-pressure loop at the bottom of the diagram. At 1 it is assumed that the exhaust and induction strokes have been completed, and the cylinder is again filled with air. The piston-cylinder system absorbs energy between 1 and 2 – this is the work needed to compress the air in the cylinder, and is provided by mechanical kinetic energy stored in the flywheel of the engine. Work output is done by the piston-cylinder combination between 2 and 4. The difference between these two increments of work is the indicated work output per cycle, and is represented by the area enclosed by the p–V loop. The adiabatic expansion is in a higher pressure range than that of the compression because the gas in the cylinder is hotter during expansion than during compression. It is for this reason that the loop has a finite area, and the net output of work during a cycle is positive.

Efficiency

Due to its high compression ratio, the diesel engine has a high efficiency, and the lack of a throttle valve means that the charge-exchange losses are fairly low, resulting in a low specific fuel consumption, especially in medium and low load situations. This makes the diesel engine very economical. Even though diesel engines have a theoretical efficiency of 75%, in practice it is much lower. In his 1893 essay *Theory and Construction of a Rational Heat Motor*, Rudolf Diesel describes that the effective efficiency of the diesel engine would be in between 43.2% and 50.4%, or maybe even greater. Modern passenger car diesel engines may have an effective efficiency of up to 43%, whilst engines in large diesel trucks, and buses can achieve peak efficiencies around 45%. However, average efficiency over a driving cycle is lower than peak efficiency. For example, it might be 37% for an engine with a peak efficiency of 44%. The highest diesel engine efficiency of up to 55% is achieved by large two-stroke watercraft diesel engines.

Major Advantages

Diesel engines have several advantages over engines operating on other principles:

- The diesel engine has the highest effective efficiency of all combustion engines.

  ◦ Diesel engines inject the fuel directly into the combustion chamber, have no intake air restrictions apart from air filters and intake plumbing and have no intake manifold vacuum to add parasitic load and pumping losses resulting from the pistons being pulled downward against intake system vacuum. Cylinder filling with atmospheric air is aided and volumetric efficiency is increased for the same reason.

  ◦ Although, the fuel efficiency (mass burned per energy produced) of a diesel engine drops at lower loads, it doesn't drop quite as fast as that of a typical petrol or turbine engine.

Bus powered by biodiesel.

- Diesel engines can combust a huge variety of fuels, including several fuel oils, that have advantages over fuels such as petrol. These advantages include:

  ○ Low fuel costs, as fuel oils are relatively cheap.

  ○ Good lubrication properties.

  ○ High energy density.

  ○ Low risk of catching fire, as they do not form a flammable vapour.

  ○ Biodiesel is an easily synthesised, non-petroleum-based fuel (through transesterification) which can run directly in many diesel engines, while gasoline engines either need adaptation to run synthetic fuels or else use them as an additive to gasoline (e.g., ethanol added to gasohol).

- Diesel engines have a very good exhaust-emission behaviour. The exhaust contains minimal amounts of carbon monoxide and hydrocarbons. Direct injected diesel engines emit approximately as much nitrogen oxide as Otto cycle engines. Swirl chamber and precombustion chamber injected engines, however, emit approximately 50% less nitrogen oxide than Otto cycle engines when running under full load. Compared with Otto cycle engines, diesel engines emit 10 times less pollutants and 3 times less carbon dioxide.

- They have no high voltage electrical ignition system, resulting in high reliability and easy adaptation to damp environments. The absence of coils, spark plug wires, etc., also eliminates a source of radio frequency emissions which can interfere with navigation and communication equipment, which is especially important in marine and aircraft applications, and for preventing interference with radio telescopes. (For this reason, only diesel-powered vehicles are allowed in parts of the American National Radio Quiet Zone.)

- Diesel engines can accept super- or turbocharging pressure without any natural limit, constrained only by the design and operating limits of engine components, such as pressure, speed and load. This is unlike petrol engines, which inevitably suffer detonation at higher pressure if engine tuning and/or fuel octane adjustments are not made to compensate.

Fuel Injection

Diesel engines rely on internal mixture formation, which means that they require a fuel injection system. The fuel is injected directly into the combustion chamber, which can be either a segmented combustion chamber or an unsegmented combustion chamber. Fuel injection with the latter is referred to as *direct injection* (DI), whilst injection into the former is called *indirect injection* (IDI). In diesel engine terminology, indirect injection does not mean fuel injection into the inlet manifold or anywhere else outside the cylinder or combustion chamber: in fact, the definition of the diesel engine excludes such injection methods. For creating the fuel pressure, diesel engines usually have an injection pump. There are several different types of injection pumps and methods for creating a fine air-fuel mixture. Over the years many different injection methods have been used. These can be described as the following:

- Air blast: Where the fuel is blown into the cylinder by a blast of air.

- Solid fuel / hydraulic injection: Where the fuel is pushed through a spring loaded valve / injector to produce a combustible mist.

- Mechanical unit injector: Where the injector is directly operated by a cam and fuel quantity is controlled by a rack or lever.

- Mechanical electronic unit injector: Where the injector is operated by a cam and fuel quantity is controlled electronically.

- Common rail mechanical injection: Where fuel is at high pressure in a common rail and controlled by mechanical means.

- Common rail electronic injection: Where fuel is at high pressure in a common rail and controlled electronically.

Torque Controlling

Due to the way diesel engines work, a vital component of all diesel engines is a mechanical or electronic governor which regulates the torque of the engine and thus idling speed and maximum speed by controlling the rate of fuel delivery. This means a change of λ_v. Unlike Otto-cycle engines, incoming air is not throttled. Mechanically governed fuel injection systems are driven by the engine's gear train. These systems use a combination of springs and weights to control fuel delivery relative to both load and speed. Modern electronically controlled diesel engines control fuel delivery by use of an electronic control module (ECM) or electronic control unit (ECU). The ECM/ECU receives an engine speed signal, as well as other operating parameters such as intake manifold pressure and fuel temperature, from a sensor and controls the amount of fuel and start of injection timing through actuators to maximise power and efficiency and minimise emissions. Controlling the timing of the start of injection of fuel into the cylinder is a key to minimizing emissions, and maximizing fuel economy (efficiency), of the engine. The timing is measured in degrees of crank angle of the piston before top dead centre. For example, if the ECM/ECU initiates fuel injection when the piston is 10° before TDC, the start of injection, or timing, is said to be 10° before TDC. Optimal timing will depend on the engine design as well as its speed and load.

Types of Fuel Injection

Air-blast Injection

Diesel's original engine injected fuel with the assistance of compressed air, which atomised the fuel and forced it into the engine through a nozzle (a similar principle to an aerosol spray). The nozzle opening was closed by a pin valve lifted by the camshaft to initiate the fuel injection before top dead centre (TDC). This is called an air-blast injection. Driving the compressor used some power but the efficiency was better than the efficiency of any other combustion engine at that time. Also, air-blast injection made engines very clunky and heavy and did not allow for quick load alteration, thus rendering it unusable for road vehicles.

Typical early 20th century air-blast injected diesel engine, rated at 59 kW.

Indirect injection

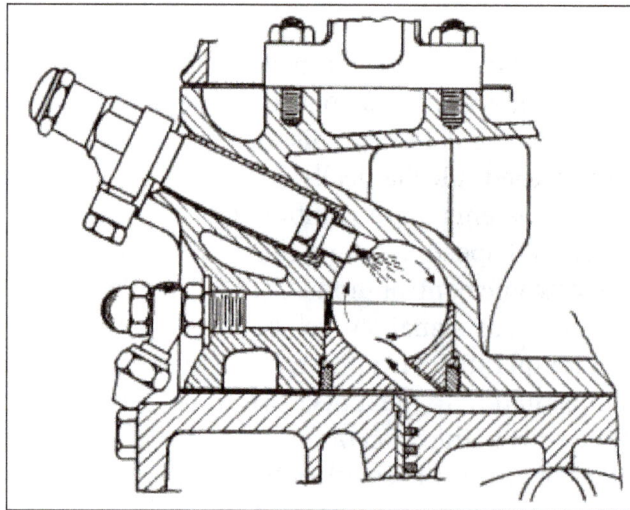

Ricardo Comet indirect injection chamber.

An indirect diesel injection system (IDI) engine delivers fuel into a small chamber called a swirl chamber, precombustion chamber, pre chamber or ante-chamber, which is connected to the cylinder by a narrow air passage. Generally the goal of the pre chamber is to create increased turbulence for better air / fuel mixing. This system also allows for a smoother, quieter running engine, and because fuel mixing is assisted by turbulence, injector pressures can be lower. Most IDI systems use a single orifice injector. The pre-chamber has the disadvantage of lowering efficiency due to increased heat loss to the engine's cooling system, restricting the combustion burn, thus reducing the efficiency by 5–10%. IDI engines are also more difficult to start and usually require the use of glow plugs. IDI engines may be cheaper to build but generally require a higher compression ratio than the DI counterpart. IDI also makes it easier to produce smooth, quieter running engines with a simple mechanical injection system since exact injection timing is not as critical. Most modern automotive engines are DI which have the benefits of greater efficiency and easier starting;

however, IDI engines can still be found in the many ATV and small diesel applications. Indirect injected diesel engines use pintle-type fuel injectiors.

Helix-controlled Direct Injection

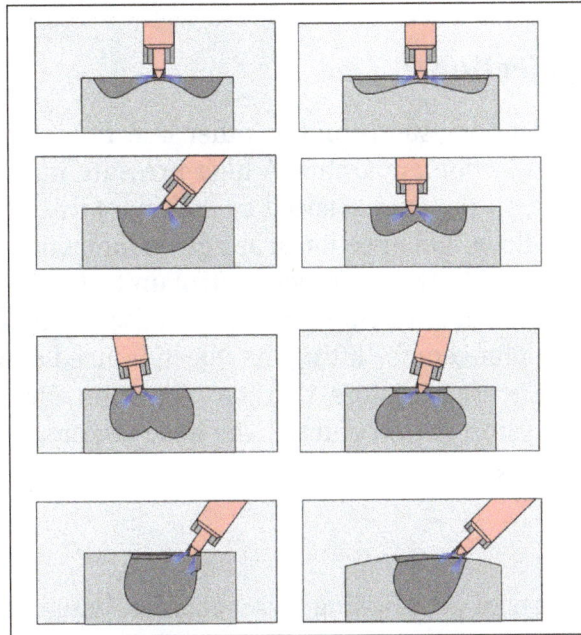

Different types of piston bowls.

Direct injection Diesel engines inject fuel directly into the cylinder. Usually there is a combustion cup in the top of the piston where the fuel is sprayed. Many different methods of injection can be used. Usually, an engine with helix-controlled mechanic direct injection has either an inline or a distributor injection pump. For each engine cylinder, the corresponding plunger in the fuel pump measures out the correct amount of fuel and determines the timing of each injection. These engines use injectors that are very precise spring-loaded valves that open and close at a specific fuel pressure. Separate high-pressure fuel lines connect the fuel pump with each cylinder. Fuel volume for each single combustion is controlled by a slanted groove in the plunger which rotates only a few degrees releasing the pressure and is controlled by a mechanical governor, consisting of weights rotating at engine speed constrained by springs and a lever. The injectors are held open by the fuel pressure. On high-speed engines the plunger pumps are together in one unit. The length of fuel lines from the pump to each injector is normally the same for each cylinder in order to obtain the same pressure delay. Direct injected diesel engines usually use orifice-type fuel injectors.

Electronic control of the fuel injection transformed the direct injection engine by allowing much greater control over the combustion.

Unit Direct Injection

Unit direct injection, also known as Pumpe-Düse (*pump-nozzle*), is a high pressure fuel injection system that injects fuel directly into the cylinder of the engine. In this system the injector and the pump are combined into one unit positioned over each cylinder controlled by the camshaft. Each cylinder has its own unit eliminating the high-pressure fuel lines, achieving a more consistent injection. Under full load, the injection pressure can reach up to 220 MPa. Unit injection systems used to dominate the commercial diesel engine market, but due to higher requirements of the flexibility of the injection system, they have been rendered obsolete by the more advanced common-rail-system.

Common Rail Direct Injection

Common rail (CR) direct injection systems, unlike other injection systems, do not have a combined pressure creation and injection apparatus. A high-pressure injection pump creates a constant pressure, not depending on the engine speed or fuel mass injected. A buffer, the so-called rail, saves this pressure. This allows fuel injection at any given moment, even multiple injections in a very short amount of time. The Electronic Diesel Control unit (EDC) controls both rail pressure and injections depending on several different parameters of the engine. The injectors of older CR systems have solenoid-driven plungers for lifting the injection needle, whilst newer CR injectors use plungers driven by piezoelectric actuators, that have fewer moving mass and therefore allow even more injections in a very short period of time. The injection pressure of modern CR systems ranges from 140 MPa to 270 MPa.

Types

There are several different ways of categorising diesel engines, based on different design characteristics:

By Power Output

- Small <188 kW (252 hp),
- Medium 188–750 kW,
- Large >750 kW.

By Cylinder Bore

- Passenger car engines: 75–100 mm,
- Lorry and commercial vehicle engines: 90–170 mm,
- High-performance high-speed engines: 165v280 mm,
- Medium-speed engines: 240–620 mm,
- Low-speed two-stroke engines: 260–900 mm.

By Number of Strokes

- Four-stroke cycle,
- Two-stroke cycle.

By Piston and Connecting Rod

- Crosshead piston,
- Double-acting piston,
- Opposed piston,
- Trunk piston.

By Cylinder Arrangement

Regular cylinder configurations such as straight (inline), V, and boxer (flat) configurations can be used for diesel engines. The inline-six-cylinder design is the most prolific in light- to medium-duty engines, though inline-four engines are also common. Small-capacity engines (generally considered to be those below five litres in capacity) are generally four- or six-cylinder types, with the four-cylinder being the most common type found in automotive uses. The V configuration used to be common for commercial vehicles, but it has been abandoned in favour of the inline configuration.

By Engine Speeds

Günter Mau categorises diesel engines by their rotational speeds into three groups:

- High-speed engines (> 1,000 rpm),
- Medium-speed engines (300–1,000 rpm), and
- Slow-speed engines (< 300 rpm).

High-speed Engines

High-speed engines are used to power trucks (lorries), buses, tractors, cars, yachts, compressors, pumps and small electrical generators. As of 2018, most high-speed engines have direct injection. Many modern engines, particularly in on-highway applications, have common rail direct injection. On bigger ships, high-speed diesel engines are often used for powering electric generators. The highest power output of high-speed diesel engines is approximately 5 MW.

Medium-speed Engines

Medium-speed engines are used in large electrical generators, ship propulsion and mechanical drive applications such as large compressors or pumps. Medium speed diesel engines operate on either diesel fuel or heavy fuel oil by direct injection in the same manner as low-speed engines. Usually, they are four-stroke engines with trunk pistons.

The power output of medium-speed diesel engines can be as high as 21,870 kW, with the effective efficiency being around 47–48%. Most larger medium-speed engines are started with compressed air direct on pistons, using an air distributor, as opposed to a pneumatic starting motor acting on the flywheel, which tends to be used for smaller engines.

Medium-speed engines intended for marine applications are usually used to power (ro-ro) ferries, passenger ships or small freight ships. Using medium-speed engines reduces the cost of smaller ships and increases their transport capacity. In addition to that, a single ship can use two smaller engines instead of one big engine, which increases the ship's safety.

Low-speed Engines

The MAN B&W 5S50MC 5-cylinder, 2-stroke, low-speed marine diesel engine.
This particular engine is found aboard a 29,000 tonne chemical carrier.

Low-speed diesel engines are usually very large in size and mostly used to power ships. There are two different types of low-speed engines that are commonly used: Two-stroke engines with a crosshead, and four-stroke engines with a regular trunk-piston. Two-stroke engines have a limited rotational frequency and their charge exchange is more difficult, which means that they are usually bigger than four-stroke engines and used to directly power a ship's propeller. Four-stroke engines on ships are usually used to power an electric generator. An electric motor powers the propeller. Both types are usually very undersquare. Low-speed diesel engines (as used in ships and other applications where overall engine weight is relatively unimportant) often have an effective efficiency of up to 55%. Like medium-speed engines, low-speed engines are started with compressed air, and they use heavy oil as their primary fuel.

Two-stroke Engines

Two-stroke diesel engines use only two strokes instead of four strokes for a complete engine cycle. Filling the cylinder with air and compressing it takes place in one stroke, and the power and exhaust strokes are combined. The compression in a two-stroke diesel engine is similar to the compression that takes place in a four-stroke diesel engine: As the piston passes through bottom centre and starts upward, compression commences, culminating in fuel injection and ignition. Instead of a full set of valves, two-stroke diesel engines have simple intake ports, and exhaust ports (or exhaust valves). When the piston approaches bottom dead centre, both the intake and the exhaust ports are "open", which means that there is atmospheric pressure inside the cylinder. Therefore, some sort of pump is required to blow the air into the cylinder and the combustion gasses into the exhaust. This process is called *scavenging*. The pressure required is approximately 10−30 kPa.

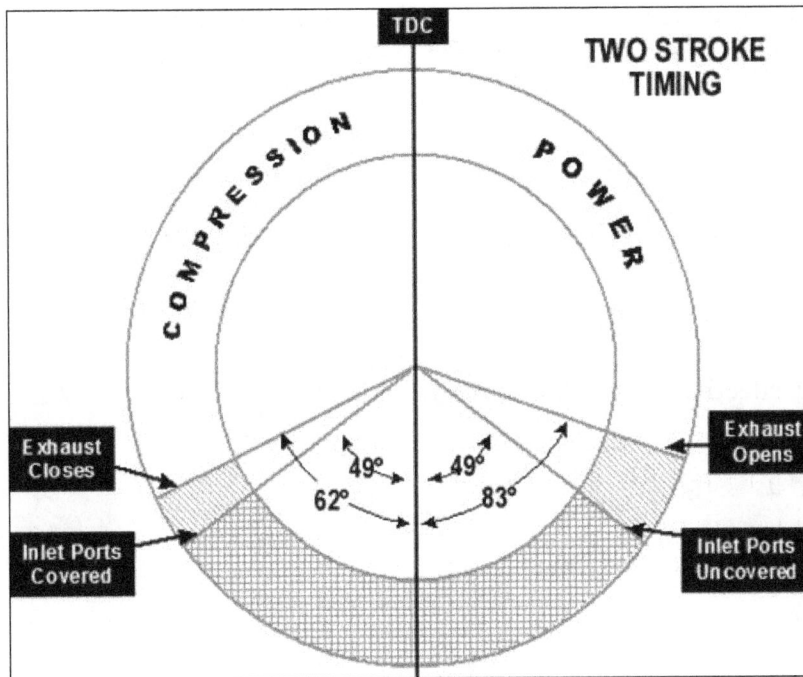

Detroit Diesel timing.

Scavenging

In general, there are three types of scavenging possible:

- Uniflow scavenging

- Crossflow scavenging

- Reverse flow scavenging

Crossflow scavenging is incomplete and limits the stroke, yet some manufacturers used it. Reverse flow scavenging is a very simple way of scavenging, and it was popular amongst manufacturers until the early 1980s. Uniflow scavenging is more complicated to make but allows the highest fuel efficiency; since the early 1980s, manufacturers such as MAN and Sulzer have switched to this system. It is standard for modern marine two-stroke diesel engines.

Dual-fuel Diesel Engines

So-called dual-fuel diesel engines or gas diesel engines burn two different types of fuel *simultaneously*, for instance, a gaseous fuel and diesel engine fuel. The diesel engine fuel auto-ignites due to compression ignition, and then ignites the gaseous fuel. Such engines do not require any type of spark ignition and operate similar to regular diesel engines.

Diesel Engine Particularities

Torque and Power

Torque is a force applied to a lever at a right angle multiplied by the lever length. This means that the torque an engine produces depends on the displacement of the engine and the force that the gas pressure inside the cylinder applies to the piston, commonly referred to as effective piston pressure:

$$M = p_e \cdot V_h \cdot \pi^{-1} \cdot i^{-1}$$

where, M:Torque [N·m]; p_e:Effective piston pressure [kN·m⁻²]; v_h: Displacement [dm³]; i: Strokes [either 2 or 4].

Example:

- Engine A: effective piston pressure=570 kN·m⁻², displacement= 2.2 dm³, strokes= 4, torque= 100 N·m,

$$570 \cdot 2.2 \cdot \pi^{-1} \cdot 4^{-1} \approx 100$$

Power is the quotient of work and time:

$$P = 2\pi n M$$

where, P: Power [W]; M: Torque [N·m]; n: Time (crankshaft speed) [s⁻¹]

which means:

$$P = 2\pi \cdot n_1 \cdot M \cdot 60^{-1}$$

where, P: Power [W]; M: Torque [N·m]; n_1: Time (crankshaft speed) [min⁻¹]

Example:

- Engine A: Power≈ 44,000 W, torque= 100 N·m, time= 4200 min⁻¹

$$44,000 \approx 2 \cdot \pi \cdot 4200 \cdot 100 \cdot 60^{-1}$$

- Engine B: Power≈ 44,000 W, torque= 260 N·m, time= 1600 min⁻¹

$$44,000 \approx 2 \cdot \pi \cdot 1600 \cdot 260 \cdot 60^{-1}$$

This means, that increasing either torque or time will result in an increase in power. As the maximum rotational frequency of the diesel engine's crankshaft is usually in between 3500–5000 min⁻¹ due to diesel principle limitations, the torque of the diesel engine must be great to achieve a high power, or, in other words, as the diesel engine cannot use a lot of time for achieving a certain amount of power, it has to perform more work (=produce more torque).

Mass

The average diesel engine has a poorer power-to-mass ratio than the Otto engine. This is because the diesel must operate at lower engine speeds. Due to the higher operating pressure inside the combustion chamber, which increases the forces on the parts due to inertial forces, the diesel engine needs heavier, stronger parts capable of resisting these forces, which results in an overall greater engine mass.

Emissions

As diesel engines burn a mixture of fuel and air, the exhaust therefore contains substances that consist of the same chemical elements, as fuel and air. The main elements of air are nitrogen (N_2) and oxygen (O_2), fuel consists of hydrogen (H_2) and carbon (C). Burning the fuel will result in the final stage of oxidation. An *ideal diesel engine*, (a hypothetical model that we use as an example), running on an ideal air-fuel mixture, produces an exhaust that consists of carbon dioxide (CO_2), water (H_2O), nitrogen (N_2), and the remaining oxygen (O_2). The combustion process in a real engine differs from an ideal engine's combustion process, and due to incomplete combustion, the exhaust contains additional substances, most notably, carbon monoxide (CO), diesel particulate matter (PM), and due to dissociation, nitrogen oxide (NO_x).

When diesel engines burn their fuel with high oxygen levels, this results in high combustion temperatures and higher efficiency, and particulate matter tends to burn, but the amount of NO_x pollution tends to increase. NO_x pollution can be reduced by recirculating a portion of an engine's exhaust gas back to the engine cylinders, which reduces the oxygen quantity, causing a reduction of combustion temperature, and resulting in fewer NO_x. To further reduce NO_x emissions, lean NO_x traps (LNTs) and SCR-catalysts can be used. Lean NO_x traps adsorb the nitrogen oxide and "trap" it. Once the LNT is full, it has to be "regenerated" using hydrocarbons. This is achieved by using a very rich air-fuel mixture, resulting in incomplete combustion. An SCR-catalyst converts nitrogen oxide using urea, which is injected into the exhaust stream, and catalytically converts the NO_x into nitrogen (N_2) and water (H_2O). Compared with an Otto engine, the diesel engine produces approximately the same amount of NO_x, but some older diesel engines may have an exhaust that contains up to 50% less NO_x. However, Otto engines, unlike diesel engines, can use a three-way-catalyst, that converts most of the NO_x.

Diesel engine exhaust composition		
Species	Mass percentage	Volume percentage
Nitrogen (N_2)	75.2%	72.1%
Oxygen (O_2)	15%	0.7%
Carbon dioxide (CO_2)	7.1%	12.3%
Water (H_2O)	2.6%	13.8%
Carbon monoxide (CO)	0.043%	0.09%
Nitrogen oxide (NO_x)	0.034%	0.13%
Hydrocarbons (HC)	0.005%	0.09%
Aldehyde	0.001%	(n/a)
Particulate matter (Sulfate + solid substances)	0.008%	0.0008%

Noise

The distinctive noise of a diesel engine is variably called diesel clatter, diesel nailing, or diesel knock. Diesel clatter is caused largely by the way the fuel ignites; the sudden ignition of the diesel fuel when injected into the combustion chamber causes a pressure wave, resulting in an audible "knock". Engine designers can reduce diesel clatter through: indirect injection; pilot or pre-injection; injection timing; injection rate; compression ratio; turbo boost; and exhaust gas recirculation (EGR). Common rail diesel injection systems permit multiple injection events as an aid to noise reduction. Therefore, newer diesel engines do not knock anymore. Diesel fuels with a higher cetane rating are more likely to ignite and hence reduce diesel clatter.

Cold Weather Starting

In general, diesel engines do not require any starting aid. In cold weather however, some diesel engines can be difficult to start and may need preheating depending on the combustion chamber design. The minimum starting temperature that allows starting without pre-heating is 40 °C for precombustion chamber engines, 20 °C for swirl chamber engines, and 0 °C for direct injected engines. Smaller engines with a displacement of less than 1 litre per cylinder usually have glowplugs, whilst larger heavy-duty engines have flame-start systems.

In the past, a wider variety of cold-start methods were used. Some engines, such as Detroit Diesel engines used a system to introduce small amounts of ether into the inlet manifold to start combustion. Instead of glowplugs, some diesel engines are equipped with starting aid systems that change valve timing. The simplest way this can be done is with a decompression lever. Activating the decompression lever locks the outlet valves in a slight down position, resulting in the engine not having any compression and thus allowing for turning the crankshaft over without resistance. When the crankshaft reaches a higher speed, flipping the decompression lever back into its normal position will abruptly re-activate the outlet valves, resulting in compression – the flywheel's mass moment of inertia then starts the engine. Other diesel engines, such as the precombustion chamber engine XII Jv 170/240 made by Ganz & Co., have a valve timing changing system that is operated by adjusting the inlet valve camshaft, moving it into a slight "late" position. This will make the inlet valves open with a delay, forcing the inlet air to heat up when entering the combustion chamber.

Supercharging and Turbocharging

Turbocharged 1980s passenger car diesel engine with wastegate turbocharger and without intercooler (BMW M21).

As the diesel engine relies on manipulation of λ_v for torque controlling and speed regulation, the intake air mass does not have to precisely match the injected fuel mass (which would be $\lambda = 1$).

diesel engines are thus ideally suited for supercharging and turbocharging. An additional advantage of the diesel engine is the lack of fuel during the compression stroke. In diesel engines, the fuel is injected near top dead centre (TDC), when the piston is near its highest position. The fuel then ignites due to compression heat. Pre-ignition, caused by the artificial turbocharger compression increase during the compression stroke, cannot occur.

Many diesels are therefore turbocharged and some are both turbocharged and supercharged. A turbocharged engine can produce more power than a naturally aspirated engine of the same configuration. A supercharger is powered mechanically by the engine's crankshaft, while a turbocharger is powered by the engine exhaust. Turbocharging can improve the fuel economy of diesel engines by recovering waste heat from the exhaust, increasing the excess air factor, and increasing the ratio of engine output to friction losses. Adding an intercooler to a turbocharged engine further increases engine performance by cooling down the air-mass and thus allowing more air-mass per volume.

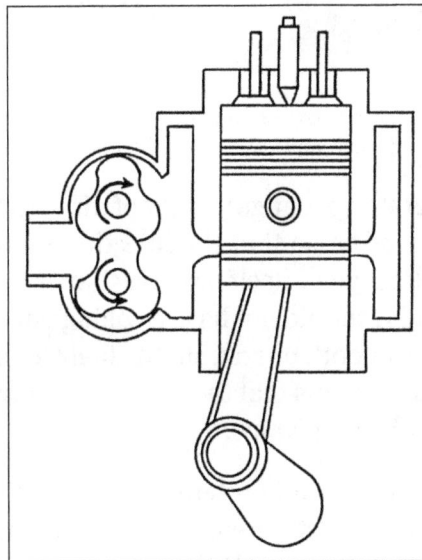

Two stroke diesel engine with Roots blower, typical of Detroit Diesel and some Electro-Motive Diesel Engines.

A two-stroke engine does not have a discrete exhaust and intake stroke and thus is incapable of self-aspiration. Therefore, all two-stroke diesel engines must be fitted with a blower or some form of compressor to charge the cylinders with air and assist in dispersing exhaust gases, a process referred to as scavenging. Roots-type superchargers were used for ship engines until the mid-1950s, since 1955 they have been widely replaced by turbochargers. Usually, a two-stroke ship diesel engine has a single-stage turbocharger with a turbine that has an axial inflow and a radial outflow.

Fuel and Fluid Characteristics

In diesel engines, a mechanical injector system vaporises the fuel directly into the combustion chamber (as opposed to a Venturi jet in a carburetor, or a fuel injector in a manifold injection system vaporising fuel into the intake manifold or intake runners as in a petrol engine). This *forced vaporisation* means that less-volatile fuels can be used. More crucially, because only air is inducted into the cylinder in a diesel engine, the compression ratio can be much higher as there is no risk of pre-ignition provided the injection process is accurately timed. This means that cylinder temperatures are much higher in a diesel engine than a petrol engine, allowing less volatile fuels to be used.

The MAN 630's M-System diesel engine is a petrol engine (designed to run on NATO F 46/F 50 petrol), but it also runs on jet fuel, (NATO F 40/F 44), kerosine, (NATO F 58), and diesel engine fuel (NATO F 54/F 75).

Therefore, diesel engines can operate on a huge variety of different fuels. In general, fuel for diesel engines should have a proper viscosity, so that the injection pump can pump the fuel to the injection nozzles without causing damage to itself or corrosion of the fuel line. At injection, the fuel should form a good fuel spray, and it should not have a coking effect upon the injection nozzles. To ensure proper engine starting and smooth operation, the fuel should be willing to ignite and hence not cause a high ignition delay, (this means that the fuel should have a high cetane number). Diesel fuel should also have a high lower heating value.

Inline mechanical injector pumps generally tolerate poor-quality or bio-fuels better than distributor-type pumps. Also, indirect injection engines generally run more satisfactorily on fuels with a high ignition delay (for instance, petrol) than direct injection engines. This is partly because an indirect injection engine has a much greater 'swirl' effect, improving vaporisation and combustion of fuel, and because (in the case of vegetable oil-type fuels) lipid depositions can condense on the cylinder walls of a direct-injection engine if combustion temperatures are too low (such as starting the engine from cold). Direct-injected engines with an MAN centre sphere combustion chamber rely on fuel condensing on the combustion chamber walls. The fuel starts vaporising only after ignition sets in, and it burns relatively smoothly. Therefore, such engines also tolerate fuels with poor ignition delay characteristics, and, in general, they can operate on petrol rated 86 RON.

Fuel Types

In his 1893 work Theory and Construction of a Rational Heat Motor, Rudolf Diesel considers using coal dust as fuel for the diesel engine. However, Diesel just *considered* using coal dust (as well as liquid fuels and gas); his actual engine was designed to operate on petroleum, which was soon replaced with regular petrol and kerosine for further testing purposes, as petroleum proved to be too viscous. In addition to kerosine and petrol, Diesel's engine could also operate on ligroin.

Before diesel engine fuel was standardised, fuels such as petrol, kerosine, gas oil, vegetable oil and mineral oil, as well as mixtures of these fuels, were used. Typical fuels specifically intended to be used for diesel engines were petroleum distillates and coal-tar distillates such as the following; these fuels have specific lower heating values of:

- Diesel oil: 10,200 kcal·kg^{-1} (42.7 MJ·kg^{-1}) up to 10,250 kcal·kg^{-1} (42.9 MJ·kg^{-1}).

- Heating oil: 10,000 kcal·kg^{-1} (41.8 MJ·kg^{-1}) up to 10,200 kcal·kg^{-1} (42.7 MJ·kg^{-1}).

- Coal-tar creosote: 9,150 kcal·kg^{-1} (38.3 MJ·kg^{-1}) up to 9,250 kcal·kg^{-1} (38.7 MJ·kg^{-1}).

- Kerosine: up to 10,400 kcal·kg^{-1} (43.5 MJ·kg^{-1}).

The first diesel fuel standards were the DIN 51601, VTL 9140-001, and NATO F 54, which appeared after World War II. The modern European EN 590 diesel fuel standard was established in May 1993; the modern version of the NATO F 54 standard is mostly identical with it. The DIN 51628 biodiesel standard was rendered obsolete by the 2009 version of the EN 590; FAME biodiesel conforms to the EN 14214 standard. Watercraft diesel engines usually operate on diesel engine fuel that conforms to the ISO 8217 standard (Bunker C). Also, some diesel engines can operate on gasses (such as LNG).

Modern Diesel Fuel Properties

Modern diesel fuel properties		
	EN 590 (as of 2009)	EN 14214 (as of 2010)
Ignition performance	≥ 51 CN	≥ 51 CN
Density at 15 °C	820–845 kg·m^{-3}	860–900 kg·m^{-3}
Sulphur content	≤10 mg·kg^{-1}	≤10 mg·kg^{-1}
Water content	≤200 mg·kg^{-1}	≤500 mg·kg^{-1}
Lubricity	460 µm	460 µm
Viscosity at 40 °C	2.0–4.5 mm²·s^{-1}	3.5–5.0 mm²·s^{-1}
FAME content	≤7.0%	≥96.5%
Molar H/C ratio	–	1.69
Lower heating value	–	37.1 MJ·kg^{-1}

Gelling

DIN 51601 diesel fuel was prone to *waxing* or *gelling* in cold weather; both are terms for the solidification of diesel oil into a partially crystalline state. The crystals build up in the fuel system (especially in fuel filters), eventually starving the engine of fuel and causing it to stop running. Low-output electric heaters in fuel tanks and around fuel lines were used to solve this problem. Also, most engines have a *spill return* system, by which any excess fuel from the injector pump and injectors is returned to the fuel tank. Once the engine has warmed, returning warm fuel prevents waxing in the tank. Some manufacturers, such as BMW, recommended fuelling diesel cars with petrol to prevent the fuel from gelling when the temperatures dropped below –15 °C.

Safety

Fuel Flammability

Diesel fuel is less flammable than petrol, because its flash point is 55 °C, leading to a lower risk of fire caused by fuel in a vehicle equipped with a diesel engine.

Diesel fuel can create an explosive air/vapour mix under the right conditions. However, compared with petrol, it is less prone due to its lower vapour pressure, which is an indication of evaporation rate. The Material Safety Data Sheet for ultra-low sulfur diesel fuel indicates a vapour explosion hazard for diesel fuel indoors, outdoors, or in sewers.

Cancer

Diesel exhaust has been classified as an IARC Group 1 carcinogen. It causes lung cancer and is associated with an increased risk for bladder cancer.

Applications

The characteristics of diesel have different advantages for different applications:

Passenger Cars

Diesel engines have long been popular in bigger cars and have been used in smaller cars such as superminis in Europe since the 1980s. They were popular in larger cars earlier, as the weight and cost penalties were less noticeable. Smooth operation as well as high low end torque are deemed important for passenger cars and small commercial vehicles. The introduction of electronically controlled fuel injection significantly improved the smooth torque generation, and starting in the early 1990s, car manufacturers began offering their high-end luxury vehicles with diesel engines. Passenger car diesel engines usually have between three and ten cylinders, and a displacement ranging from 0.8 to 5.0 litres. Modern powerplants are usually turbocharged and have direct injection.

Diesel engines do not suffer from intake-air throttling, resulting in very low fuel consumption especially at low partial load (for instance: driving at city speeds). One fifth of all passenger cars worldwide have diesel engines, with many of them being in Europe, where approximately 47% of all passenger cars are diesel-powered. Daimler-Benz in conjunction with Robert Bosch GmbH produced diesel-powered passenger cars starting in 1936. The popularity of diesel-powered passenger cars in markets such as India, South Korea and Japan is increasing.

Commercial Vehicles and Lorries

In 1893, Rudolf Diesel suggested that the diesel engine could possibly power 'wagons' (lorries). The first lorries with diesel engines were brought to market in 1924.

Modern diesel engines for lorries have to be both extremely reliable and very fuel efficient. Common-rail direct injection, turbocharging and four valves per cylinder are standard. Displacements range from 4.5 to 15.5 litres, with power-to-mass ratios of 2.5–3.5 kg·kW^{-1} for heavy duty and

2.0–3.0 kg·kW⁻¹ for medium duty engines. V6 and V8 engines used to be common, due to the relatively low engine mass the V configuration provides. Recently, the V configuration has been abandoned in favour of straight engines. These engines are usually straight-6 for heavy and medium duties and straight-4 for medium duty. Their undersquare design causes lower overall piston speeds which results in increased lifespan of up to 1,200,000 km. Compared with 1970s diesel engines, the expected lifespan of modern lorry diesel engines has more than doubled.

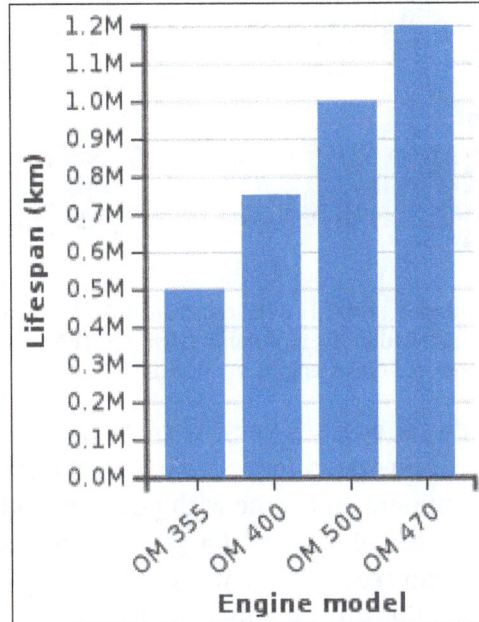

Lifespan of Mercedes-Benz diesel engines.

Railroad Rolling Stock

Diesel engines for locomotives are built for continuous operation and may require the ability to use poor quality fuel in some circumstances. Some locomotives use two-stroke diesel engines. Diesel engines have eclipsed steam engines as the prime mover on all non-electrified railroads in the industrialised world. The first diesel locomotives appeared in 1913, and diesel multiple units soon after. Many modern diesel locomotives are actually diesel-electric locomotives: the diesel engine is used to power an electric generator that in turn powers electric traction motors with no mechanical connection between diesel engine and traction. While electric locomotives have replaced the diesel locomotive for some passenger traffic in Europe and Asia, diesel is still today very popular for cargo-hauling freight trains and on tracks where electrification is not feasible.

In the 1940s, road vehicle diesel engines with power outputs of 150–200 PS (110–147 kW) were considered reasonable for DMUs. Commonly, regular truck powerplants were used. The height of these engines had to be less than 1,000 mm to allow underfloor installation. Usually, the engine was mated with a pneumatically operated mechanical gearbox, due to the low size, mass, and production costs of this design. Some DMUs used hydraulic torque converters instead. Diesel-electric transmission was not suitable for such small engines. In the 1930s, the Deutsche Reichsbahn standardised its first DMU engine. It was a 30.3 litre, 12-cylinder boxer unit, producing 275 PS (202 kW). Several German manufacturers produced engines according to this standard.

Watercraft

One of the eight-cylinder 3200 I.H.P. Harland and Wolff – Burmeister & Wain
diesel engines installed in the motorship *Glenapp*. This was the highest
powered diesel engine yet installed in a ship.

The requirements for marine diesel engines vary, depending on the application. For military use and medium-size boats, medium-speed four-stroke diesel engines are most suitable. These engines usually have up to 24 cylinders and come with power outputs in the one-digit Megawatt region. Small boats may use lorry diesel engines. Large ships use extremely efficient, low-speed two-stroke diesel engines. They can reach efficiencies of up to 55%. Unlike most regular diesel engines, two-stroke watercraft engines use highly viscous fuel oil. Submarines are usually diesel-electric.

The first diesel engines for ships were made by A. B. Diesels Motorer Stockholm in 1903. These engines were three-cylinder units of 120 PS (88 kW) and four-cylinder units of 180 PS (132 kW) and used for Russian ships. In World War I, especially submarine diesel engine development advanced quickly. By the end of the War, double acting piston two-stroke engines with up to 12,200 PS (9 MW) had been made for marine use.

Aviation

Diesel engines had been used in aircraft before World War II, for instance, in the rigid airship *LZ 129 Hindenburg*, which was powered by four Daimler-Benz DB 602 diesel engines, or in several Junkers aircraft, which had Jumo 205 engines installed. Until the late 1970s, there has not been any applications of the diesel engine in aircraft. In 1978, Karl H. Bergey argued that "the likelihood of a general aviation diesel in the near future is remote." In recent years, diesel engines have found use in unmanned aircraft (UAV), due to their reliability, durability, and low fuel consumption. In early 2019, AOPA reported, that a diesel engine model for general aviation aircraft is "approaching the finish line."

Non-road Diesel Engines

Non-road diesel engines are commonly used for construction equipment. Fuel efficiency, reliability and ease of maintenance are very important for such engines, whilst high power output and

quiet operation are negligible. Therefore, mechanically controlled fuel injection and air-cooling are still very common. The common power outputs of non-road diesel engines vary a lot, with the smallest units starting at 3 kW, and the most powerful engines being heavy duty lorry engines.

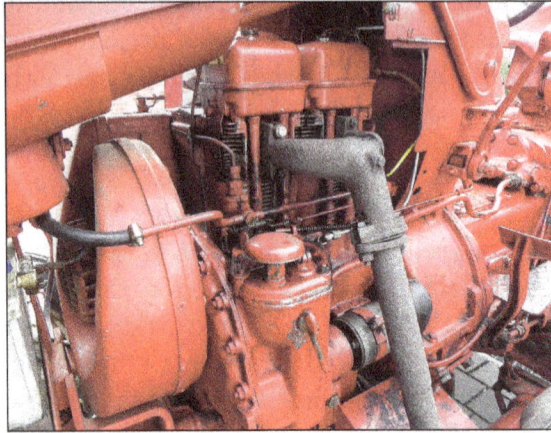

Air-cooled diesel engine of a 1959 Porsche 218.

Stationary Diesel Engines

Stationary diesel engines are commonly used for electricity generation, but also for powering refrigerator compressors, or other types of compressors or pumps. Usually, these engines run permanently, either with mostly partial load, or intermittently, with full load. Stationary diesel engines powering electric generators that put out an alternating current, usually operate with alternating load, but fixed rotational frequency. This is due to the mains' fixed frequency of either 50 Hz (Europe), or 60 Hz (United States). The engine's crankshaft rotational frequency is chosen so that the mains' frequency is a multiple of it. For practical reasons, this results in crankshaft rotational frequencies of either 25 Hz (1500 per minute) or 30 Hz (1800 per minute).

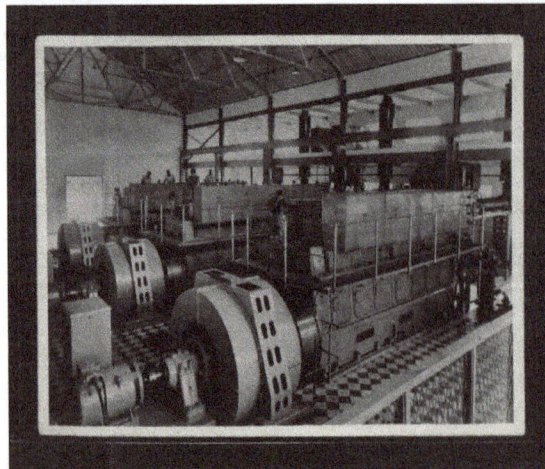

Three English Electric 7SRL diesel-alternator sets being installed at the Saateni Power Station.

Low Heat Rejection Engines

A special class of prototype internal combustion piston engines has been developed over several decades with the goal of improving efficiency by reducing heat loss. These engines are variously

called adiabatic engines; due to better approximation of adiabatic expansion; low heat rejection engines, or high temperature engines. They are generally piston engines with combustion chamber parts lined with ceramic thermal barrier coatings. Some make use of pistons and other parts made of titanium which has a low thermal conductivity and density. Some designs are able to eliminate the use of a cooling system and associated parasitic losses altogether. Developing lubricants able to withstand the higher temperatures involved has been a major barrier to commercialization.

Future Developments

In mid-2010s literature, main development goals for future diesel engines are described as improvements of exhaust emissions, reduction of fuel consumption, and increase of lifespan. It is said that the diesel engine, especially the diesel engine for commercial vehicles, will remain the most important vehicle powerplant until the mid-2030s. Editors assume that the complexity of the diesel engine will increase further. Some editors expect a future convergency of diesel and Otto engines' operating principles due to Otto engine development steps made towards homogeneous charge compression ignition.

Spark-ignition Engine

It is an internal combustion engine in which the ignition of the air-fuel mixture takes place by the spark. The spark is generated with the help of spark plug. Since in this engine, the spark is responsible for the ignition of the fuel, it is named as spark ignition engine (SI engine). This engine uses petrol as a fuel for its working. It works on the principle of otto cycle. The fuel in this engine is injected through carburetor during suction stroke. The compression ratio of this engine is usually 6 to 10. It has light weight engine and used in light duty vehicles like motorcycle, cars etc.

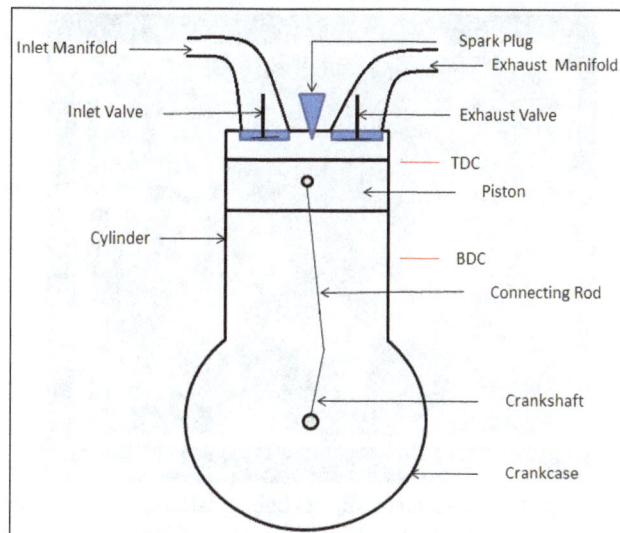

Main Components of spark ignition engine.

The main components of spark ignition engine are as follows:

- Inlet Valve: Air-fuel mixture enters into cylinder through inlet valve.

- Exhaust Valve: The burnt or exhaust gases produced in the power stroke escapes out through exhaust valve.

- Spark Plug: It produces spark at the end of the compression stroke, which ignites the compressed air-fuel mixture.

- Cylinder: It is a hollow cylinder in which the piston reciprocates.

- Piston: It is moving part of the engine that performs reciprocating motion and transmits the power generated during power stroke to the crankshaft through connecting rod.

- Connecting Rod: It is that part of the engine which connects the piston to the crankshaft.

- Crankshaft: It is used to convert the reciprocating motion of the engine into rotary motion.

Working

- Suction Stroke: Air-fuel mixture enters into the cylinder.

- Compression Stroke: Compression of air fuel mixture takes place.

- Power Stroke: Combustion of Fuel and Power Generation.

- Exhaust Stroke: Escaping of burnt gases out of the engine.

Suction Stroke

In this stroke, the piston moves downward and the air-fuel mixture from the carburetor enters into the cylinder through inlet valve. During this stroke inlet valve opens and exhaust valve remains closed.

Compression Stroke

In this stroke, the piston moves upward and compresses the air-fuel mixture. The compression strokes completes as the piston moves at TDC. During this stroke Both the inlet and exhaust valve remains closed.

Power Stroke

At the end of compression stroke, a spark is produced by the spark plug. This spark ignites the air-fuel mixture and combustion takes place in the combustion chamber. Due to combustion, a very high thrust force is generated which pushes the piston downward rapidly and makes the crankshaft to rotate. This stroke is called as power stroke because we get power in it. Both inlet and exhaust valve remains closed in this stroke.

Exhaust Stroke

In this stroke, the piston moves upward and burnt or exhaust gases produced in the power stroke escapes out of the cylinder through exhaust valve. In this stroke, the exhaust valve gets open and inlet valve remains closed.

After the completion of exhaust stroke, again all the four stroke repeats itself. The most commonly used spark ignition engine are of two stroke engine and four stroke engine. In two stroke engine we have inlet and exhaust port instead of valve.

The position of inlet and exhaust valve and operation performed during all the four strokes of the SI engine are given in the table below.

S.no	Stroke	Inlet Valve	Exhaust Valve	Operation Performed
1.	Suction stroke	Open	Closed	Suction of fuel
2.	Compression stroke	Closed	Closed	Compression of fuel
3.	Power stroke	Closed	Closed	Combustion of fuel
4.	Exhaust stroke	Closed	Open	Escaping of burnt gases

Application

The spark ignition engine are used in automobiles (motorcycle, Scooters, cars etc), aircraft, motor-boats and in small engines such as chainsaws, lawn mowers etc.

Jet Engine

Jet engine is any of a class of internal-combustion engines that propel aircraft by means of the rearward discharge of a jet of fluid, usually hot exhaust gases generated by burning fuel with air drawn in from the atmosphere.

Characteristics

The prime mover of virtually all jet engines is a gas turbine. Variously called the core, gas producer, gasifier, or gas generator, the gas turbine converts the energy derived from the combustion of a liquid hydrocarbon fuel to mechanical energy in the form of a high-pressure, high-temperature airstream. This energy is then harnessed by what is termed the propulsor (e.g., airplane propeller and helicopter rotor) to generate a thrust with which to propel the aircraft.

Principles of Operation

The Prime Mover

The gas turbine operates on the Brayton cycle in which the working fluid is a continuous flow of air ingested into the engine's inlet. The air is first compressed by a turbocompressor to a pressure ratio of typically 10 to 40 times the pressure of the inlet airstream. It then flows into a combustion chamber, where a steady stream of the hydrocarbon fuel, in the form of liquid spray droplets and vapour or both, is introduced and burned at approximately constant pressure. This gives rise to a continuous stream of high-pressure combustion products whose average temperature is typically from 980 to 1,540 °C or higher. This stream of gases flows through a turbine, which is linked by a torque shaft to the compressor and which extracts energy from the gas stream to drive the compressor. Because heat has been added to the working fluid at high pressure, the gas stream that exits from the gas generator after having been expanded through the turbine contains a considerable amount of surplus energy—i.e., gas horsepower—by virtue of its high pressure, high temperature, and high velocity, which may be harnessed for propulsion purposes.

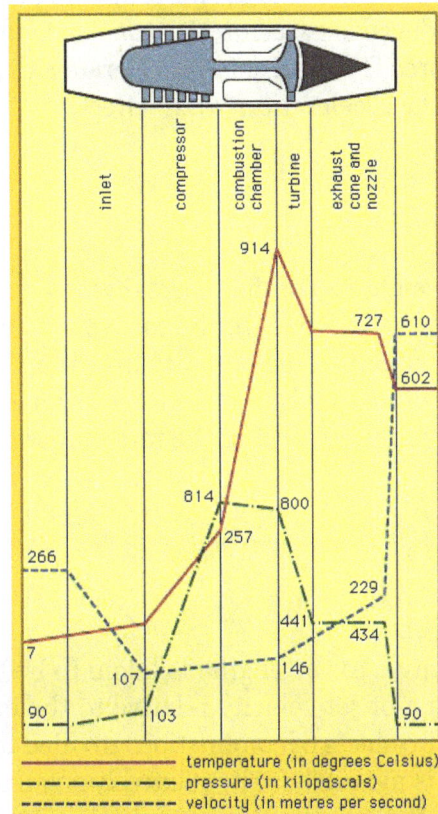

Cross section of a turbojet and graph of typical operating conditions for its working fluid.

The heat released by burning a typical jet fuel in air is approximately 43,370 kilojoules per kilogram (18,650 British thermal units per pound) of fuel. If this process were 100 percent efficient, it would then produce a gas power for every unit of fuel flow of 7.45 horsepower/(pounds per hour), or 12 kilowatts/(kg per hour). In actual fact, certain practical thermodynamic limitations, which are a function of the peak gas temperature achieved in the cycle, restrict the efficiency of the process to about 40 percent of this ideal value. The peak pressure achieved in the cycle also affects the efficiency of energy generation. This implies that the lower limit of specific fuel consumption (SFC) for an engine producing gas horsepower is 0.336 (pound per hour)/horsepower, or 0.207 (kg per hour)/kilowatt. In actual practice, the SFC is even higher than this lower limit because of inefficiencies, losses, and leakages in the individual components of the prime mover.

Because weight and volume are at a premium in the overall design of an aircraft and because the power plant represents a large fraction of any aircraft's total weight and volume, these parameters must be minimized in the engine design. The airflow that passes through an engine is a representative measure of the engine's cross-sectional area and hence its weight and volume. Therefore, an important figure of merit for the prime mover is its specific power—the amount of power that it generates per unit of airflow. This quantity is a very strong function of the peak gas temperature in the core at the discharge of the combustion chamber. Modern engines generate from 150 to 250 horsepower/(pound per second), or 247 to 411 kilowatts/(kg per second).

The Propulsor

The gas horsepower generated by the prime mover in the form of hot, high-pressure gas is used to drive the propulsor, enabling it to generate thrust for propelling or lifting the aircraft. The principle on which such a thrust is produced is based on Newton's second law of motion. This law generalizes the observation that the force (F) required to accelerate a discrete mass (m) is proportional to the product of that mass and the acceleration (a). In effect,

$$F = ma = \frac{wa}{g},$$

where the mass is taken as the weight (w) of the object divided by the acceleration due to gravity (g) at the place where the object was weighed. In the case of a jet engine, one is generally dealing with the acceleration of a steady stream of air rather than with a discrete mass. Here, the equivalent statement of the second law of motion is that the force (F) required to increase the velocity of a stream of fluid is proportional to the product of the rate of mass flow (M) of the stream and the change in velocity of the stream,

$$F = M\left(V_j - V_0\right) = \frac{W\left(V_j - V_0\right)}{g},$$

where the inlet velocity (V_o) relative to the engine is taken to be the flight velocity and the discharge velocity (V_j) is the exhaust or jet velocity relative to the engine. W is the rate of weight flow of working fluid (i.e., air or products of combustion) divided by the acceleration of gravity in the place where the weight flow is measured. The relatively small effect of the weight flow of fuel in creating a difference between the weight flow of the inlet and exhaust streams is intentionally disregarded.

One thereby infers that the components of a propulsor must exert a force F on the stream of air flowing through the propulsor if this device accelerates the airstream from the flight velocity V_o to the discharge velocity V_j. The reaction to that force F is ultimately transmitted by the mounts of the propulsor to the aircraft as propulsive thrust.

There are two general approaches to converting gas horsepower to propulsive thrust. In one, a second turbine (i.e., a low-pressure, or power, turbine) may be introduced into the engine flow path to extract additional mechanical power from the available gas horsepower. This mechanical power may then be used to drive an external propulsor, such as an airplane propeller or helicopter rotor. In this case, the thrust is developed in the propulsor as it energizes and accelerates the airflow through the propulsor—i.e., an airstream separate from that flowing through the prime mover.

In the second approach, the high-energy stream delivered by the prime mover may be fed directly to a jet nozzle, which accelerates the gas stream to a very high velocity as it leaves the engine, as is typified by the turbojet. In this case, the thrust is developed in the components of the prime mover as they energize the gas stream.

In other types of engines, such as the turbofan, thrust is generated by both approaches: A major part of the thrust is derived from the fan, which is powered by a low-pressure turbine and which energizes and accelerates the bypass stream. The remaining part of the total thrust is derived from the core stream, which is exhausted through a jet nozzle.

Just as the prime mover is an imperfect device for converting the heat of fuel combustion to gas horsepower, so the propulsor is an imperfect device for converting the gas horsepower to propulsive thrust. There is generally a great deal of energy left in the high-temperature, high-velocity jet stream exiting from the propulsor that is not fully exploited for propulsion. The efficiency of a propulsor, propulsive efficiency η_p, is the portion of the available energy that is usefully applied in propelling the aircraft compared to the total energy of the jet stream. For the simple but representative case of the discharge airflow equal to the inlet gas flow, it is found that:

$$\eta_p = \frac{2V_0}{V_j + V_0}$$

Although the jet velocity V_j must be larger than the aircraft velocity V_0 to generate useful thrust, a large jet velocity that exceeds flight speed by a substantial margin can be very detrimental to propulsive efficiency. Maximum propulsive efficiency is approached when the jet velocity is almost equal to (but, of necessity, slightly higher than) the flight speed. This fundamental fact has given rise to a large variety of jet engines, each designed to generate a specific range of jet velocities that matches the range of flight speeds of the aircraft that it is supposed to power.

The net assessment of the efficiency of a jet engine is the measurement of its rate of fuel consumption per unit of thrust generated (e.g., in terms of pounds, or kilograms, per hour of fuel consumed per pounds, or kilograms, of thrust generated). There is no simple generalization of the value of specific fuel consumption of a thrust engine. It is not only a strong function of the prime mover's efficiency (and hence its pressure ratio and peak-cycle temperature) but also of the propulsive efficiency of the propulsor (and hence of the engine type). It also is a strong function of the aircraft flight speed and the ambient temperature (which is in turn a strong function of altitude, season, and latitude).

Basic Engine Types

Achieving a high propulsive efficiency for a jet engine is dependent on designing it so that the exiting jet velocity is not greatly in excess of the flight speed. At the same time, the amount of thrust generated is proportional to that very same velocity excess that must be minimized. This set of restrictive requirements has led to the evolution of a large number of specialized variations of the basic turbojet engine, each tailored to achieve a balance of good fuel efficiency, low weight, and compact size for duty in some band of the flight speed–altitude–mission spectrum. There are two major general features characteristic of all the different engine types, however. First, in order to achieve a high propulsive efficiency, the jet velocity, or the velocity of the gas stream exiting the propulsor, is matched to the flight speed of the aircraft—slow aircraft have engines with low jet velocities and fast aircraft have engines with high jet velocities. Second, as a result of designing the jet velocity to match the flight speed, the size of the propulsor varies inversely with the flight speed of the aircraft—slow aircraft have very large propulsors, as, for example, the helicopter rotor—and the relative size of the propulsor decreases with increasing design flight speed—turboprop propellers are relatively small and turbofan fans even smaller.

Although the turbojet is the simplest jet engine and was invented and flown first among all the engine types, it seems useful to examine the entire spectrum of engines in the order of the flight-speed band in which they serve, starting with the slowest—namely, the turboshaft engine, which powers helicopters.

Turboshaft Engines

The helicopter is designed to operate for substantial periods of time hovering at zero flight speed. Even in forward flight, helicopters rarely exceed 240 kilometres per hour or a Mach number of 0.22. (The Mach number is the ratio of the velocity of the aircraft to the speed of sound.) The principal propulsor is the helicopter rotor, which is driven by one or more turboshaft engines in all modern helicopters of large size. As was previously noted, the propulsor is designed to give a very low discharge or jet velocity and is by the same token very large for a given size aircraft when compared to the propulsors of higher-speed aircraft. The prime mover of a helicopter is a core engine whose gas horsepower is extracted by a power turbine, which then drives the helicopter rotor via a speed-reducing (and combining) gearbox. The power turbine is usually located on a spool separate from the gas generator; thus its rotative speed and that of the helicopter rotor which it drives are independent of the rotative speed of the gas generator. This allows the rotor speed to be varied or kept constant independently of the gas-generator speed, which must be varied to modulate the amount of power generated.

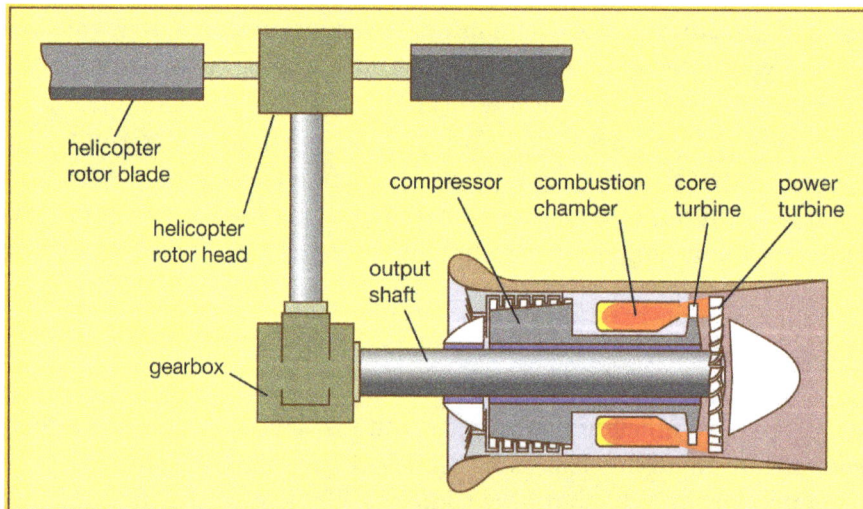

Turboshaft engine driving a helicopter rotor as propulsor.

Turboprops, Propfans and Unducted Fan Engines

The turboprop is the power plant that occupies the next band of flight speeds in the flight spectrum, from a Mach number of 0.2 to 0.7. The propulsor is a propeller with a somewhat higher discharge, or jet velocity, than that of the helicopter rotor to match the flight speed, and it has a proportionately smaller area than the latter for a similarly sized aircraft. The prime mover is a turboshaft engine very similar to the one that drives a helicopter rotor except for a different gearbox; designed to provide a somewhat higher rotative speed for the propeller, which turns faster than the helicopter rotor having a much larger diameter. The control mode of the turboprop also is somewhat different from that of a helicopter's turboshaft engine. In a helicopter the pilot calls for power by manipulating the pitch of the rotor blades (a greater pitch taking a bigger "bite" of air and so demanding more power to maintain rotative speed). The engine's control responds by increasing fuel to the engine to maintain output shaft speed. In a turboprop the pilot calls for power by selecting fuel flow to the prime mover. The propeller control responds by varying propeller pitch to attain a greater "pull" while maintaining a preselected propeller rotative speed.

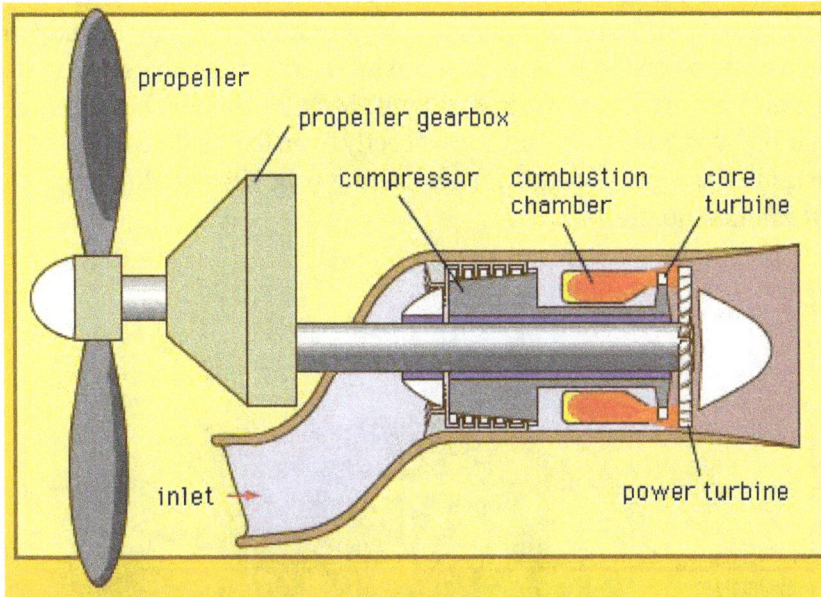

Turboprop engine driving a single rotation propeller as propulsor; tractor arrangement.

A recent trend in turboprop design has been the evolution of propellers for efficient operation at transonic flight speeds (those approximating the speed of sound), much higher than previously achieved—up to Mach numbers of 0.85. This usually involves a higher disk loading (i.e., a higher discharge velocity from the propeller) in order to permit the use of a smaller diameter propeller. This trend has been accompanied by an increase in the number of blades in the propeller (from 6 to 12 instead of the more common 2 to 4 blades in lower-speed propellers). The blades are scimitar-shaped, with swept-back leading edges at the blade tips to accommodate the large Mach numbers encountered by the propeller tip at high rotative and flight speeds. Such high-speed propulsors are called propfans.

Another variation of the propulsor involves the application of two concentric propellers on the same centreline, driven by the same prime mover through a gearbox that causes each propeller to rotate in a direction opposite the other. Such counter-rotating propellers are capable of significantly higher propulsive efficiency and higher disk loading than conventional propellers.

In most turboprop installations the prime mover is mounted on the wing, and the plane of the propeller is forward of the prime mover (the so-called tractor layout). Modern high-speed aircraft may find it more advantageous to mount the engine more toward the rear of the aircraft, with the plane of the propeller aft of the engine. These arrangements are referred to as "pusher" layouts. A recently developed engine layout, identified as the unducted fan, provides a set of very high-efficiency counter-rotating propeller blades, each blade mounted on one of either of two sets of counter-rotating low-pressure turbine stages and achieving all the advantages of the arrangement without the use of a gearbox.

Medium-bypass Turbofans, High-bypass Turbofans and Ultrahigh-bypass Engines

Moving up in the spectrum of flight speeds to the transonic regime—Mach numbers from 0.75 to 0.9—the most common engine configurations are turbofan engines, such as those shown in figures. In a turbofan, only a part of the gas horsepower generated by the core is extracted to drive a propulsor, which usually consists of a single low-pressure-ratio, shrouded turbocompression stage. The fan is generally placed in front of the core inlet so that the air entering the core first passes through the fan and is partially compressed by it. Most of the air, however, bypasses the core (hence the designation bypass stream) and goes directly to an exhaust nozzle. The core stream, with some modest fraction of the gas horsepower remaining (not extracted to drive the fan) proceeds directly to its own exhaust nozzle.

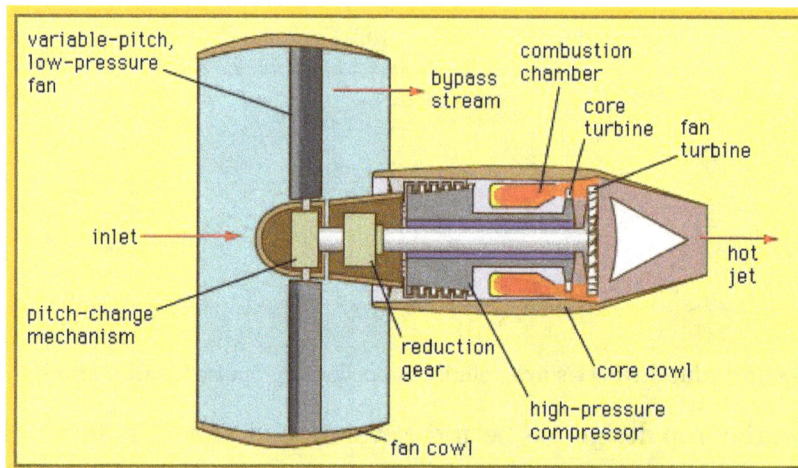

Ultrahigh-bypass engine (UBE) with geared fan and variable-pitch blading for thrust reversal.

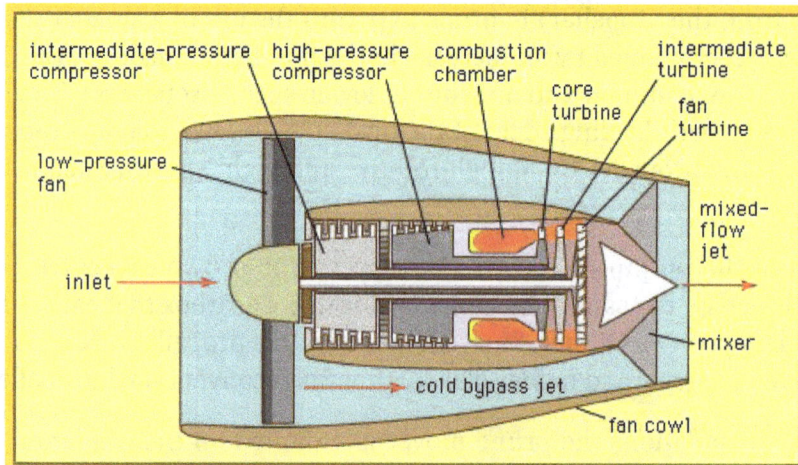

High-bypass turbofan with two-spool core and mixed-flow jet.

A key parameter for classifying the turbofan is its bypass ratio, defined as the ratio of the mass flow rate of the bypass stream to the mass flow rate entering the core. Since the highest propulsion efficiencies are obtained by the engines with the highest bypass ratios, one would expect to find all engines of that design in this flight speed regime. (Some of the variation derives from historical evolution.) In actuality, however, one finds engines with a broad spectrum of bypass ratios, including medium-bypass engines (with bypass ratios from 2 to 4), high-bypass engines (with bypass ratios from 5 to 8), and ultrahigh-bypass engines, so-called UBEs (with bypass ratios from 9 to 15 or higher). A whole generation of low- and medium-bypass engines has completely supplanted the first generation of aircraft powered by (zero-bypass) turbojet engines. Moreover, that generation was itself supplanted by a third generation of medium- and high-bypass turbofan engines. There are several other reasons why engines with less than the highest bypass ratios hypothetically achievable are still in use. Very high bypass ratios involve the use of fans with very large diameters, which in turn entail very heavy components; this increases the difficulty of installing the engine on aircraft and maintaining sufficient ground clearance. In addition, the weight and complexity of the apparatus required to reverse the direction of the bypass stream (to achieve thrust reversal in order to shorten the aircraft's landing roll) also increases with the bypass ratio. The long-term trend, however, is definitely toward higher and higher bypass ratios.

There are several unique features and ancillary devices found in turbofan engines. Ultrahigh-bypass engines may have a gearbox between the drive turbine and the fan to simplify the design of the small-diameter turbine (with the attendant high rotative speed) without compromising the performance of the very large-diameter fan (with the attendant low rotative speed). Variable-pitch fan blades are generally required for thrust reversal in such ultrahigh-bypass fans, while in medium- and high-bypass engines the thrust reversing is usually accomplished by introducing blocker doors into the bypass stream. In high- and medium-bypass turbofans, a small but significant improvement in propulsive efficiency can be achieved by mixing the airstream of the hot core and cold bypass streams before the total airstream enters a single jet nozzle.

Low-bypass Turbofans and Turbojets

In the next higher regime of aircraft flight speed, the low supersonic range from Mach numbers above 1 up to 2 or 3, one finds the application of the simple turbojet (with no bypass stream) and the low-bypass turbofan engine (with a bypass ratio up to 2).

Although the low-bypass turbofan has the same general appearance as a turbofan with a larger bypass ratio, certain special features are unique to low-bypass engines. The lower total flow in the fan generally involves a higher fan pressure ratio (for equivalent amounts of energy available from the drive turbine), and so such a fan usually has more than one (i.e., two or three) turbocompressor stages. Engines designed to operate at the low supersonic range generally have insufficient thrust in other flight regimes or modes where they must operate for short durations, as, for instance, acceleration through transonic speed, takeoff from high-altitude airports under conditions of extremely high temperatures and high gross weight, or combat maneuvers at high supersonic flight speed. Rather than installing a larger engine to meet these requirements, it is more effective to add an afterburner to a turbofan engine as a means of thrust augmentation. The afterburner is a secondary combustion system that operates in the exhaust stream of the engine before the stream is introduced into the exhaust nozzle. Such a device is not as fuel-efficient as the main turbofan section of the engine because heat addition occurs at a lower pressure than in the main burner. The afterburner, however, is relatively simple and lightweight, since it does not contain any rotating machinery. For the same reason, it may be operated to a much higher discharge temperature (typically 1,760 °C), so that it is capable of augmenting the thrust of the turbofan by as much as 50 percent.

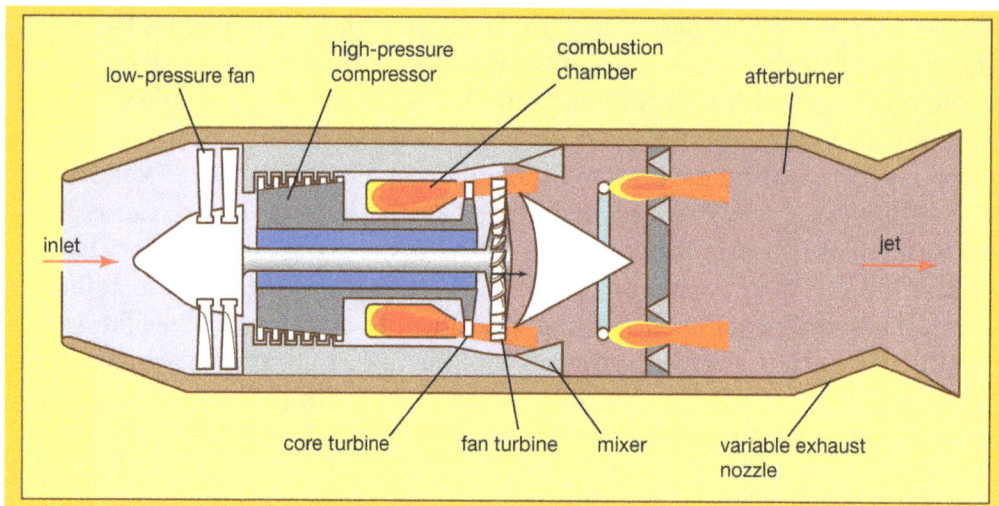

Low-bypass turbofan with afterburner.

The afterburner in a turbofan usually requires a mixer for mixing the relatively cool bypass air with the hot core stream; the cooler air is otherwise difficult to burn in the low-pressure environment of an afterburner. Also, in both the turbojet and the turbofan with an afterburner, the exhaust nozzle must have a variable throat area to accommodate the large variations in volumetric flow rate between the very hot exhaust stream from the operating afterburner and the cooler airstream discharged from the engine when the afterburner is not in use. Engines intended for supersonic flight generally have a much lower compression-pressure ratio than higher-bypass machines intended for subsonic or transonic operation. A major contributor to this tendency is the additional pressure ratio developed in the engine's inlet as it slows down or diffuses the very high-speed airstream that is ingested as the engine's working fluid—the ram effect. At transonic flight speed this pressure ratio is almost 2:1, so that the engine's compressor may be built to provide that much less pressure where peak pressure is otherwise limiting.

Early generations of jet-propelled aircraft in this low supersonic flight regime were powered by turbojet engines, but subsequent generations built for the same flight regime have largely been equipped with low-bypass turbofans. This substitution of engine type was undertaken primarily because such aircraft expend a great deal of their fuel at subsonic flight speed (e.g., in takeoff, climb, loiter, acceleration, approach, and landing), where the turbofan provides an advantage in propulsive efficiency.

Ramjets and Supersonic Combustion Ramjets

Ram pressure plays an increasingly important role in the thermodynamic cycle of power and thrust generation of the jet engine at supersonic flight speeds. For flight speeds above Mach 2.5 or 3, the ram-pressure ratio becomes so high that a turbocompressor is no longer necessary for efficient thrust generation. Indeed, the pressure ratio eventually rises to such high values that the associated high ram temperatures make it difficult or impossible to place high-speed rotating machinery in the flow path without prohibitive amounts of cooling provision. This combination of circumstances gives rise to the ramjet, a jet engine in which the pressure increase is attributable only to the ram effect of the high flight speed; no turbomachinery is involved, and the main thrust producer is an afterburner.

Ramjets are lightweight and simple power plants, making them ideal candidates for supersonic flight vehicles that are launched from other flight vehicles at extremely high speed. They are less suitable for use in vehicles that must be sufficiently self-powered for subsonic takeoff, climb, and acceleration to supersonic flight speed; the subsonic ram pressure is insufficient to produce any reasonable amount of thrust, and so alternative propulsion devices must be provided.

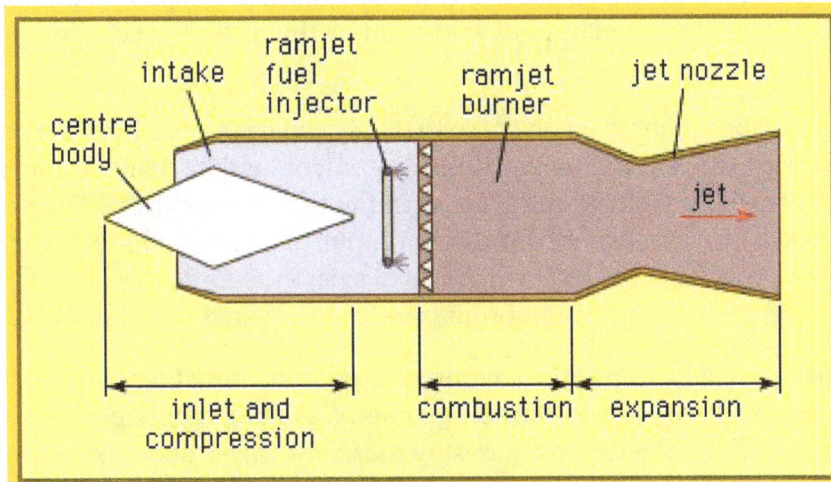

Arrangement of a ramjet.

In the flight regime of Mach 4 or 5, it is usually efficient to decelerate the inlet airstream to subsonic velocity before it enters the combustion system. At still higher Mach numbers, such deceleration becomes more difficult and costly in terms of pressure losses, and it is necessary to make provision for the combustion chamber to burn its fuel in the supersonic airstream. Such specialized ramjets are called scramjets (for supersonic combustion ramjets) and are projected to be fueled by a cryogenically liquified gas (e.g., hydrogen or methane) instead of a liquid hydrocarbon. The primary reason for doing so is to exploit the greater heat release per unit weight of fuels that have a higher ratio of hydrogen to carbon atoms than ordinary fractions of petroleum even though this gain is partly negated by the higher volume per unit of heat release of those same fuels. Another incentive for employing a very cold fuel is that it may be used as a heat sink for cooling a very high-speed (and hence very hot) engine and aircraft structure. The scramjet has an unusual feature: the inlet deceleration and exhaust acceleration occur largely outside the enclosed engine inlet and exhaust ducts against external aircraft surfaces in front of and to the rear of the engine. In effect, the engine itself is little more than a sophisticated supersonic combustion chamber.

Hybrid Engine Types

It is possible to tailor an engine configuration so that the engine is well suited for operation within a given band of the flight spectrum. To have an engine that will perform well in more than one band of the flight spectrum or in more than one regime of operation, it may be necessary to configure the power plant so that it can be converted from one engine type to another by means of variable geometry built into the engine components.

Vertical and Short Takeoff and Landing (V/STOL) Propulsion Systems

Propulsion systems that provide aircraft with the capability of both vertical and conventional forward flight represent a formidable challenge to the engine designer. V/STOL aircraft have several major categories of engine arrangement. They are as follows:

1. As in a helicopter, the propulsor may consist of a rotor that is driven by one or more turboshaft engines and is installed in such a way as to provide vertical thrust. The entire aircraft must be tilted to give the thrust vector a forward component to achieve forward flight. This arrangement has certain limitations in terms of effectiveness, as borne out by the relative inefficiency of forward flight above a Mach number of 0.2.

2. The propulsors may be mounted on pivots so that they can be rotated from the position in which they give vertical thrust in a takeoff, hover, climb, descent, or landing maneuver and pivoted 90° to provide thrust for conventional forward flight (as in the tilt-rotor aircraft). The prime mover that drives the propulsor may either be tilted with the propulsor or be fixed in the wing and drive the tilting propulsor via a rotating shaft through the pivot axis. In some configurations, the entire wing of the aircraft, carrying fixed engines and propulsors, may be tilted as a single assembly.

3. The engines may be fixed in a position required to produce thrust for forward flight. Their exhaust systems, however, have built-in variable geometry, making it possible to vector the exhaust nozzle (or nozzles) or divert the exhaust gases by means of valves and auxiliary ducts to nozzles mounted in such a way as to provide vertical thrust or lift.

4. The aircraft may include two different sets of engines or propulsors (or both), fixed in position, with one set installed for forward flight and the other for vertical thrust (i.e., the lift engines).

5. The aircraft may use a convertible engine. Such an engine has a single prime mover that is arranged to drive a fan for efficient forward propulsion, to drive a shaft that turns the main helicopter rotor, or to drive both a fan and a shaft. In order to convert from horizontal to vertical flight, variable-pitch fan blades or variable-pitch stators (or both) unload the fan, thereby making mechanical power available to drive the helicopter rotor for vertical movement.

Variable-cycle Engines

For aircraft designed to fly mixed missions (i.e., at subsonic, transonic, and supersonic flight speeds) with low levels of fuel consumption, it is desirable to have an engine with the characteristics of both a high-bypass engine (for subsonic flight speed) and a low-bypass engine (for supersonic flight speed). This requirement is typical for many high-speed commercial airliners, including the Concorde, a type of supersonic transport built by the British and French that was in service from 1976 to 2003. The Concorde was capable of traveling over oceans and unpopulated land areas at supersonic cruise speeds, but it could not fly efficiently and quietly at subsonic flight speed for takeoff, ascent, cruising over populated areas, and approach and landing. This dual function is expected to be accomplished in the future by the variable-cycle engine (VCE). If the components of an engine are designed to accommodate the extreme limits of flow, pressure ratio, and other conditions involved in both high-bypass and low-bypass operation, the engine may be operated at either extreme of bypass ratio or at any bypass ratio between those extremes by means of a valve (or valves) in the bypass stream (in conjunction with a variable exhaust nozzle). When the valves are closed, they restrict the flow in the bypass stream to achieve low bypass for supersonic flight. When the valves are open, the bypass is increased to its maximum value for efficient subsonic flight.

Turboramjets

The ramjet provides a simple and efficient means of propulsion for aircraft at relatively high supersonic flight speeds. It is, however, quite inefficient at transonic flight speeds and is completely ineffective at subsonic velocities. The turboramjet has been developed to overcome this inadequacy. In this system, a turbofan engine is built into the inlet of a ramjet engine to charge the latter with a pressurized stream of air at subsonic flight speed where ram pressure is insufficient for effective ramjet operation. During supersonic flight the fan blades, if they are of variable pitch, may be feathered so that they do not interfere with the flow of ram air to the ramjet. A separate inlet to the core engine that drives the fan may be closed off so as not to expose the turbomachinery to the hostile environment of the high-temperature ram air.

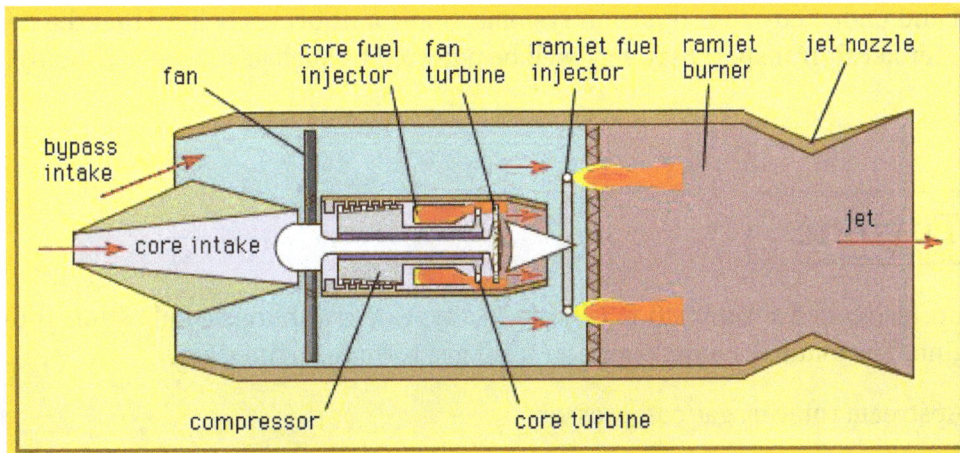

Turboramjet with air-breathing prime mover.

Another variation of the turboramjet does without the core inlet and the core compressor altogether. Instead, the aircraft carries a tank of an oxidizer, such as liquid oxygen. The oxidizer is fed into the core combustion chamber along with the fuel to support the combustion process, which generates the hot gas stream to power the turbine that drives the fan. During supersonic flight, the fan may be feathered, and a surplus of fuel may be introduced into the core combustor. The unburned fuel passes through the fan turbine and undergoes combustion in the ramjet burner when it mixes with the fresh air entering via the bypass stream from the fan.

Development of Jet Engines

Like many other inventions, jet engines were envisaged long before they became a reality. The earliest proposals were based on adaptations of piston engines and were usually heavy and complicated. The first to incorporate a turbine design was conceived as early as 1921, and the essentials of the modern turbojet were contained in a patent in 1930 by Frank Whittle in England. His design was first tested in 1937 and achieved its first flight in May 1941. In Germany, parallel but completely independent work followed issuance of a patent in 1935. It proceeded more rapidly, and the very first flight of a turbojet-powered aircraft, a Heinkel HE-178, came in August 1939. By the end of World War II these prototype aircraft had developed into a few operational turbojet squadrons in the German, British, and U.S. air forces.

In the military area, jet fighter aircraft developed rapidly and were in use during the Korean

War (1950–53), flying at speeds of 1,000 km per hour. During the next decade, they overcame the sound barrier and established normal operations up to more than twice the speed of sound (Mach 2). Bomber and transport jet aircraft were also able to reach and cruise at supersonic speeds.

The first civil jet transport, the British de Havilland Comet, flew in 1949, and regular transatlantic jet services were started in 1958 with the Comet 4 and the American Boeing 707. By 1974, more than 90 percent of hours flown throughout the world were flown by jets; the first supersonic airliner, the British-French Concorde flying at more than twice the speed of sound entered regular service in January 1976 and flew until late 2003.

During the 1980s, various major aircraft manufacturers undertook programs to develop fuel-saving propfan and unducted-fan propulsion systems. Some authorities believe that the next generation of commercial air transport may very well be powered by such advanced-technology propeller engines.

Gas Turbine

A gas turbine, also called a combustion turbine, is a type of continuous combustion, internal combustion engine. The main elements common to all gas turbine engines are:

- An upstream rotating gas compressor.
- A combustor.
- A downstream turbine on the same shaft as the compressor.

A fourth component is often used to increase efficiency (on turboprops and turbofans), to convert power into mechanical or electric form (on turboshafts and electric generators), or to achieve greater thrust-to-weight ratio (on afterburning engines).

The basic operation of the gas turbine is a Brayton cycle with air as the working fluid. Atmospheric air flows through the compressor that brings it to higher pressure. Energy is then added by spraying fuel into the air and igniting it so the combustion generates a high-temperature flow. This high-temperature high-pressure gas enters a turbine, where it expands down to the exhaust pressure, producing a shaft work output in the process. The turbine shaft work is used to drive the compressor; the energy that is not used for compressing the working fluid comes out in the exhaust gases that can be used to do external work, such as directly producing thrust in a turbojet engine, or rotating a second, independent turbine (known as a power turbine) which can be connected to a fan, propeller, or electrical generator. The purpose of the gas turbine determines the design so that the most desirable split of energy between the thrust and the shaft work is achieved. The fourth step of the Brayton cycle (cooling of the working fluid) is omitted, as gas turbines are open systems that do not use the same air again.

Gas turbines are used to power aircraft, trains, ships, electrical generators, pumps, gas compressors, and tanks.

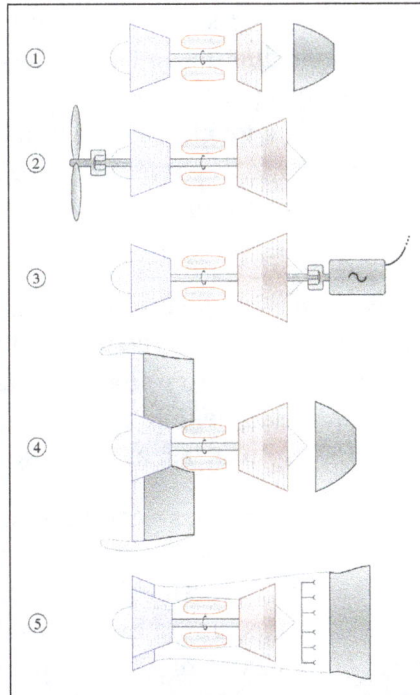

Examples of gas turbine configurations: (1) turbojet, (2) turboprop, (3) turboshaft (electric generator), (4) high-bypass turbofan, (5) low-bypass afterburning turbofan.

Theory of Operation

In an ideal gas turbine, gases undergo four thermodynamic processes: an isentropic compression, an isobaric (constant pressure) combustion, an isentropic expansion and heat rejection. Together, these make up the Brayton cycle.

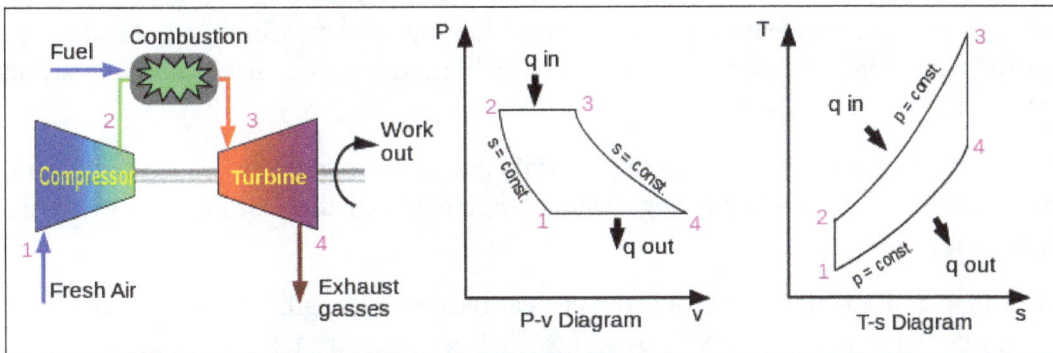

Brayton cycle.

In a real gas turbine, mechanical energy is changed irreversibly (due to internal friction and turbulence) into pressure and thermal energy when the gas is compressed (in either a centrifugal or axial compressor). Heat is added in the combustion chamber and the specific volume of the gas increases, accompanied by a slight loss in pressure. During expansion through the stator and rotor passages in the turbine, irreversible energy transformation once again occurs. Fresh air is taken in, in place of the heat rejection.

If the engine has a power turbine added to drive an industrial generator or a helicopter rotor, the exit pressure will be as close to the entry pressure as possible with only enough energy left to overcome the pressure losses in the exhaust ducting and expel the exhaust. For a turboprop engine there will be a particular balance between propeller power and jet thrust which gives the most economical operation. In a turbojet engine only enough pressure and energy is extracted from the flow to drive the compressor and other components. The remaining high-pressure gases are accelerated through a nozzle to provide a jet to propel an aircraft.

The smaller the engine, the higher the rotation rate of the shafts must be to attain the required blade tip speed. Blade-tip speed determines the maximum pressure ratios that can be obtained by the turbine and the compressor. This, in turn, limits the maximum power and efficiency that can be obtained by the engine. In order for tip speed to remain constant, if the diameter of a rotor is reduced by half, the rotational speed must double. For example, large jet engines operate around 10,000-25,000 rpm, while micro turbines spin as fast as 500,000 rpm.

Mechanically, gas turbines *can* be considerably less complex than internal combustion piston engines. Simple turbines might have one main moving part, the compressor/shaft/turbine rotor assembly, with other moving parts in the fuel system. This, in turn, can translate into price. For instance, costing 10,000 RM for materials, the Jumo 004 proved cheaper than the Junkers 213 piston engine, which was 35,000 RM, and needed only 375 hours of lower-skill labor to complete (including manufacture, assembly, and shipping), compared to 1,400 for the BMW 801. This, however, also translated into poor efficiency and reliability. More advanced gas turbines (such as those found in modern jet engines or combined cycle power plants) may have 2 or 3 shafts (spools), hundreds of compressor and turbine blades, movable stator blades, and extensive external tubing for fuel, oil and air systems; they use temperature resistant alloys, and are made with tight specifications requiring precision manufacture. All this often makes the construction of a simple gas turbine more complicated than a piston engine.

Moreover, to reach optimum performance in modern gas turbine power plants the gas needs to be prepared to exact fuel specifications. Fuel gas conditioning systems treat the natural gas to reach the exact fuel specification prior to entering the turbine in terms of pressure, temperature, gas composition, and the related wobbe-index.

The primary advantage of a gas turbine engine is its power to weight ratio. Since significant useful work can be generated by a relatively lightweight engine, gas turbines are perfectly suited for aircraft propulsion.

Thrust bearings and journal bearings are a critical part of a design. They are hydrodynamic oil bearings or oil-cooled rolling-element bearings. Foil bearings are used in some small machines such as micro turbines and also have strong potential for use in small gas turbines/auxiliary power units

Creep

A major challenge facing turbine design is reducing the creep that is induced by the high temperatures. Because of the stresses of operation, turbine materials, especially turbine blades, become damaged through these mechanisms. As temperatures are increased in an effort to improve turbine efficiency, creep becomes more significant. To limit creep, thermal coatings and superalloys with solid-solution strengthening and grain boundary strengthening are used in blade designs. Protective coatings are used to reduce the thermal damage and to limit oxidation. These coatings are often stabilized zirconium dioxide-based ceramics. Using a thermal protective coating limits the temperature exposure of the nickel superalloy. This reduces the creep mechanisms experienced in the blade. Oxidation coatings limit efficiency losses caused by a buildup on the outside of the blades, which is especially important in the high-temperature environment. The nickel-based blades are alloyed with aluminum and titanium to improve strength and creep resistance. The microstructure of these alloys is composed of different regions of the composition. A uniform dispersion of the gamma-prime phase – a combination of nickel, aluminum, and titanium – promotes the strength and creep resistance of the blade due to the microstructure. Refractory elements such as rhenium and ruthenium can be added to the alloy to improve creep strength. The addition of these elements reduces the diffusion of the gamma prime phase, thus preserving the fatigue resistance, strength, and creep resistance.

Types

Jet Engines

Typical axial-flow gas turbine turbojet, the J85, sectioned for display. Flow is left to right, multistage compressor on left, combustion chambers center, two-stage turbine on right.

Airbreathing jet engines are gas turbines optimized to produce thrust from the exhaust gases, or from ducted fans connected to the gas turbines. Jet engines that produce thrust from the direct impulse of exhaust gases are often called turbojets, whereas those that generate thrust with the addition of a ducted fan are often called turbofans or (rarely) fan-jets.

Gas turbines are also used in many liquid fuel rockets, where gas turbines are used to power a turbopump to permit the use of lightweight, low-pressure tanks, reducing the empty weight of the rocket.

Turboprop Engines

A turboprop engine is a turbine engine that drives an aircraft propeller using a reduction gear. Turboprop engines are used on small aircraft such as the general-aviation Cessna 208 Caravan and Embraer EMB 312 Tucano military trainer, medium-sized commuter aircraft such as the Bombardier Dash 8 and large aircraft such as the Airbus A400M transport and the 60-year-old Tupolev Tu-95 strategic bomber.

Aeroderivative Gas Turbines

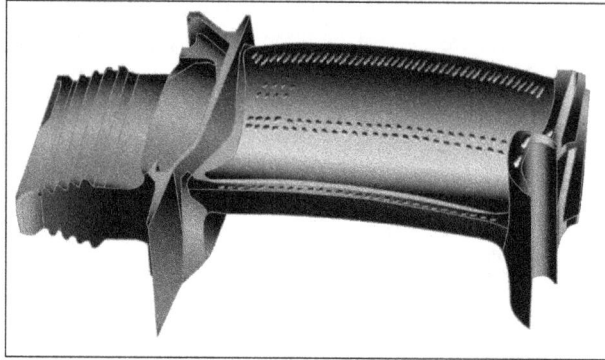

Diagram of a high-pressure film-cooled turbine blade.

Aeroderivatives are also used in electrical power generation due to their ability to be shut down and handle load changes more quickly than industrial machines. They are also used in the marine industry to reduce weight. The General Electric LM2500, General Electric LM6000, Rolls-Royce RB211 and Rolls-Royce Avon are common models of this type of machine.

Amateur Gas Turbines

Increasing numbers of gas turbines are being used or even constructed by amateurs.

In its most straightforward form, these are commercial turbines acquired through military surplus or scrapyard sales, then operated for display as part of the hobby of engine collecting. In its most extreme form, amateurs have even rebuilt engines beyond professional repair and then used them to compete for the Land Speed Record.

The simplest form of self-constructed gas turbine employs an automotive turbocharger as the core component. A combustion chamber is fabricated and plumbed between the compressor and turbine sections.

More sophisticated turbojets are also built, where their thrust and light weight are sufficient to power large model aircraft. The Schreckling design constructs the entire engine from raw materials, including the fabrication of a centrifugal compressor wheel from plywood, epoxy and wrapped carbon fibre strands.

Several small companies now manufacture small turbines and parts for the amateur. Most turbojet-powered model aircraft are now using these commercial and semi-commercial microturbines, rather than a Schreckling-like home-build.

Auxiliary Power Units

APUs are small gas turbines designed to supply auxiliary power to larger, mobile, machines such as an aircraft. They supply:

- Compressed air for air conditioning and ventilation,

- Compressed air start-up power for larger jet engines,

- Mechanical (shaft) power to a gearbox to drive shafted accessories or to start large jet engines,

- Electrical, hydraulic and other power-transmission sources to consuming devices remote from the APU.

Industrial Gas Turbines for Power Generation

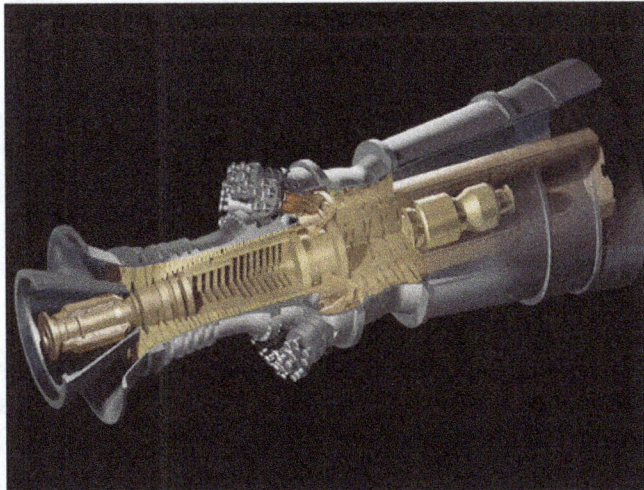

GE H series power generation gas turbine: in combined cycle configuration, its highest thermal efficiency is 62.22%.

Industrial gas turbines differ from aeronautical designs in that the frames, bearings, and blading are of heavier construction. They are also much more closely integrated with the devices they power— often an electric generator—and the secondary-energy equipment that is used to recover residual energy (largely heat).

They range in size from portable mobile plants to large, complex systems weighing more than a hundred tonnes housed in purpose-built buildings. When the gas turbine is used solely for shaft power, its thermal efficiency is about 30%. However, it may be cheaper to buy electricity than to generate it. Therefore, many engines are used in CHP (Combined Heat and Power) configurations that can be small enough to be integrated into portable container configurations.

Gas turbines can be particularly efficient when waste heat from the turbine is recovered by a heat recovery steam generator to power a conventional steam turbine in a combined cycle configuration. The 605 MW General Electric 9HA achieved a 62.22% efficiency rate with temperatures as high as 1,540 °C (2,800 °F). For 2018, GE offers its 826 MW HA at over 64% efficiency in combined cycle due to advances in additive manufacturing and combustion breakthroughs, up from 63.7% in 2017 orders and on track to achieve 65% by the early 2020s.

Aeroderivative gas turbines can also be used in combined cycles, leading to a higher efficiency, but it will not be as high as a specifically designed industrial gas turbine. They can also be run in a cogeneration configuration: the exhaust is used for space or water heating, or drives an absorption chiller for cooling the inlet air and increase the power output, technology known as Turbine Inlet Air Cooling.

Another significant advantage is their ability to be turned on and off within minutes, supplying power during peak, or unscheduled demand. Since, single cycle (gas turbine only) power plants are less efficient than combined cycle plants, they are usually used as peaking power plants, which operate anywhere from several hours per day to a few dozen hours per year—depending on the electricity demand and the generating capacity of the region. In areas with a shortage of base-load and load following power plant capacity or with low fuel costs, a gas turbine powerplant may regularly operate most hours of the day. A large single-cycle gas turbine typically produces 100 to 400 megawatts of electric power and has 35–40% thermal efficiency.

Industrial Gas Turbines for Mechanical Drive

Industrial gas turbines that are used solely for mechanical drive or used in collaboration with a recovery steam generator differ from power generating sets in that they are often smaller and feature a dual shaft design as opposed to a single shaft. The power range varies from 1 megawatt up to 50 megawatts. These engines are connected directly or via a gearbox to either a pump or compressor assembly. The majority of installations are used within the oil and gas industries. Mechanical drive applications increase efficiency by around 2%.

Oil and gas platforms require these engines to drive compressors to inject gas into the wells to force oil up via another bore, or to compress the gas for transportation. They are also often used to provide power for the platform. These platforms do not need to use the engine in collaboration with a CHP system due to getting the gas at an extremely reduced cost (often free from burn off gas). The same companies use pump sets to drive the fluids to land and across pipelines in various intervals.

Compressed Air Energy Storage

One modern development seeks to improve efficiency in another way, by separating the compressor and the turbine with a compressed air store. In a conventional turbine, up to half the generated power is used driving the compressor. In a compressed air energy storage configuration, power, perhaps from a wind farm or bought on the open market at a time of low demand and low price, is used to drive the compressor, and the compressed air released to operate the turbine when required.

Turboshaft Engines

Turboshaft engines are often used to drive compression trains (for example in gas pumping stations or natural gas liquefaction plants) and are used to power almost all modern helicopters. The primary shaft bears the compressor and the high-speed turbine (often referred to as the *Gas Generator*), while a second shaft bears the low-speed turbine (a *power turbine* or *free-wheeling turbine* on helicopters, especially, because the gas generator turbine spins separately from the power turbine). In effect the separation of the gas generator, by a fluid coupling (the hot energy-rich combustion gases), from the power turbine is analogous to an automotive transmission's fluid coupling. This arrangement is used to increase power-output flexibility with associated highly-reliable control mechanisms.

Radial Gas Turbines

In 1963, Jan Mowill initiated the development at Kongsberg Våpenfabrikk in Norway. Various successors have made good progress in the refinement of this mechanism. Owing to a configuration that keeps heat away from certain bearings, the durability of the machine is improved while the radial turbine is well matched in speed requirement.

Scale Jet Engines

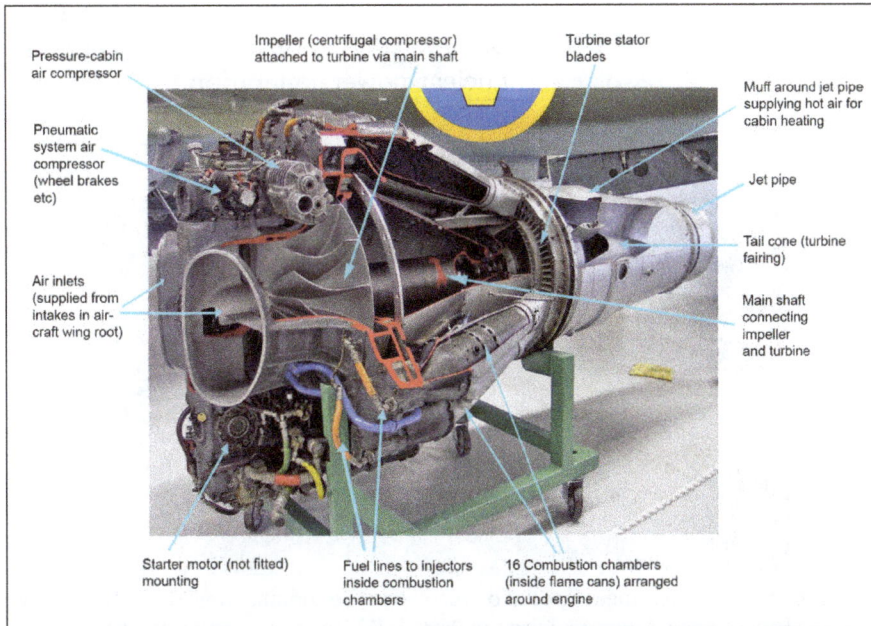

Scale jet engines are scaled down versions of this early full scale engine.

Also known as miniature gas turbines or micro-jets.

With this in mind the pioneer of modern Micro-Jets, Kurt Schreckling, produced one of the world's first Micro-Turbines, the FD3/67. This engine can produce up to 22 newtons of thrust, and can be built by most mechanically minded people with basic engineering tools, such as a metal lathe.

Microturbines

Evolved from piston engine turbochargers, aircraft APUs or small jet engines, microturbines are 25 to 500 kilowatt turbines the size of a refrigerator. Microturbines have around 15% efficiencies without a recuperator, 20 to 30% with one and they can reach 85% combined thermal-electrical efficiency in cogeneration.

External Combustion

Most gas turbines are internal combustion engines but it is also possible to manufacture an external combustion gas turbine which is, effectively, a turbine version of a hot air engine. Those systems are usually indicated as EFGT (Externally Fired Gas Turbine) or IFGT (Indirectly Fired Gas Turbine).

External combustion has been used for the purpose of using pulverized coal or finely ground biomass (such as sawdust) as a fuel. In the indirect system, a heat exchanger is used and only clean air with no combustion products travels through the power turbine. The thermal efficiency is lower in the indirect type of external combustion; however, the turbine blades are not subjected to combustion products and much lower quality (and therefore cheaper) fuels are able to be used.

When external combustion is used, it is possible to use exhaust air from the turbine as the primary combustion air. This effectively reduces global heat losses, although heat losses associated with the combustion exhaust remain inevitable.

Closed-cycle gas turbines based on helium or supercritical carbon dioxide also hold promise for use with future high temperature solar and nuclear power generation.

In Surface Vehicles

The 1967 *STP* Oil Treatment Special on display at the Indianapolis Motor Speedway Hall of Fame Museum, with the Pratt & Whitney gas turbine shown.

A 1968 Howmet TX, the only turbine-powered race car to have won a race.

Gas turbines are often used on ships, locomotives, helicopters, tanks, and to a lesser extent, on cars, buses, and motorcycles.

A key advantage of jets and turboprops for airplane propulsion - their superior performance at high altitude compared to piston engines, particularly naturally aspirated ones - is irrelevant in most automobile applications. Their power-to-weight advantage, though less critical than for aircraft, is still important.

Gas turbines offer a high-powered engine in a very small and light package. However, they are not as responsive and efficient as small piston engines over the wide range of RPMs and powers needed in vehicle applications. In series hybrid vehicles, as the driving electric motors are mechanically detached from the electricity generating engine, the responsiveness, poor performance at low speed and low efficiency at low output problems are much less important. The turbine can be run at optimum speed for its power output, and batteries and ultracapacitors can supply power as needed, with the engine cycled on and off to run it only at high efficiency. The emergence of the continuously variable transmission may also alleviate the responsiveness problem.

Turbines have historically been more expensive to produce than piston engines, though this is partly because piston engines have been mass-produced in huge quantities for decades, while small gas turbine engines are rarities; however, turbines are mass-produced in the closely related form of the turbocharger.

The turbocharger is basically a compact and simple free shaft radial gas turbine which is driven by the piston engine's exhaust gas. The centripetal turbine wheel drives a centrifugal compressor wheel through a common rotating shaft. This wheel supercharges the engine air intake to a degree that can be controlled by means of a wastegate or by dynamically modifying the turbine housing's geometry (as in a VGT turbocharger). It mainly serves as a power recovery device which converts a great deal of otherwise wasted thermal and kinetic energy into engine boost.

Turbo-compound engines (actually employed on some trucks) are fitted with blow down turbines which are similar in design and appearance to a turbocharger except for the turbine shaft being mechanically or hydraulically connected to the engine's crankshaft instead of to a centrifugal compressor, thus providing additional power instead of boost. While the turbocharger is a pressure turbine, a power recovery turbine is a velocity one.

Passenger Road Vehicles (Cars, Bikes and Buses)

A number of experiments have been conducted with gas turbine powered automobiles, the largest by Chrysler. More recently, there has been some interest in the use of turbine engines for hybrid electric cars. For instance, a consortium led by micro gas turbine company Bladon Jets has secured investment from the Technology Strategy Board to develop an Ultra Lightweight Range Extender (ULRE) for next-generation electric vehicles. The objective of the consortium, which includes luxury car maker Jaguar Land Rover and leading electrical machine company SR Drives, is to produce the world's first commercially viable - and environmentally friendly - gas turbine generator designed specifically for automotive applications.

The common turbocharger for gasoline or diesel engines is also a turbine derivative.

Concept Cars

The first serious investigation of using a gas turbine in cars was in 1946 when two engineers, Robert Kafka and Robert Engerstein of Carney Associates, a New York engineering firm, came up with the concept where a unique compact turbine engine design would provide power for a rear wheel drive car. After an article appeared in *Popular Science*, there was no further work, beyond the paper stage.

In 1950, designer F.R. Bell and Chief Engineer Maurice Wilks from British car manufacturers Rover unveiled the first car powered with a gas turbine engine. The two-seater JET1 had the engine positioned behind the seats, air intake grilles on either side of the car, and exhaust outlets on the top of the tail. During tests, the car reached top speeds of 140 km/h (87 mph), at a turbine speed of 50,000 rpm. The car ran on petrol, paraffin (kerosene) or diesel oil, but fuel consumption problems proved insurmountable for a production car. It is on display at the London Science Museum.

The 1950 Rover JET1.

A French turbine powered car, the Socema-Gregoire, was displayed at the October 1952 Paris Auto Show. It was designed by the French engineer Jean-Albert Grégoire.

GM Firebird I.

The first turbine-powered car built in the US was the GM Firebird I which began evaluations in 1953. While photos of the Firebird I may suggest that the jet turbine's thrust propelled the car like an aircraft, the turbine actually drove the rear wheels. The Firebird 1 was never meant as a commercial passenger car and was solely built for testing & evaluation as well as public relation purposes.

Starting in 1954 with a modified Plymouth, the American car manufacturer Chrysler demonstrated several prototype gas turbine-powered cars from the early 1950s through the early 1980s. Chrysler built fifty Chrysler Turbine Cars in 1963 and conducted the only consumer trial of gas turbine-powered cars. Each of their turbines employed a unique rotating recuperator, referred to as a regenerator that increased efficiency.

In 1954, FIAT unveiled a concept car with a turbine engine, called Fiat Turbina. This vehicle, looking like an aircraft with wheels, used a unique combination of both jet thrust and the engine driving the wheels. Speeds of 282 km/h (175 mph) were claimed.

Engine compartment of a Chrysler 1963 Turbine car.

The original General Motors Firebird was a series of concept cars developed for the 1953, 1956 and 1959 Motorama auto shows, powered by gas turbines.

As a result of the U.S. Clean Air Act Amendments of 1970, research was funded into developing automotive gas turbine technology. Design concepts and vehicles were conducted by Chrysler, General Motors, Ford (in collaboration with AiResearch), and American Motors (in conjunction with Williams Research). Long-term tests were conducted to evaluate comparable cost efficiency. Several AMC Hornets were powered by a small Williams regenerative gas turbine weighing 250 lb (113 kg) and producing 80 hp (60 kW; 81 PS) at 4450 rpm.

Toyota demonstrated several gas turbine powered concept cars, such as the Century gas turbine hybrid in 1975, the Sports 800 Gas Turbine Hybrid in 1979 and the GTV in 1985. No production vehicles were made. The GT24 engine was exhibited in 1977 without a vehicle.

In the early 1990s, Volvo introduced the Volvo Environmental Concept Car(ECC) which was a gas turbine powered hybrid car.

In 1993, General Motors introduced the first commercial gas turbine powered hybrid vehicle—as a limited production run of the EV-1 series hybrid. A Williams International 40 kW turbine drove an alternator which powered the battery-electric powertrain. The turbine design included a recuperator. In 2006, GM went into the EcoJet concept car project with Jay Leno.

At the 2010, Paris Motor Show Jaguar demonstrated its Jaguar C-X75 concept car. This electrically powered supercar has a top speed of 204 mph (328 km/h) and can go from 0 to 62 mph (0 to 100 km/h) in 3.4 seconds. It uses Lithium-ion batteries to power four electric motors which combine to produce 780 bhp. It will travel 68 miles (109 km) on a single charge of the batteries, and uses a pair of Bladon Micro Gas Turbines to re-charge the batteries extending the range to 560 miles (900 km).

Racing Cars

The first race car (in concept only) fitted with a turbine was in 1955 by a US Air Force group as a hobby project with a turbine loaned them by Boeing and a race car owned by Firestone Tire & Rubber company. The first race car fitted with a turbine for the goal of actual racing was by Rover and the BRM Formula One team joined forces to produce the Rover-BRM, a gas turbine powered coupe, which entered the 1963 24 Hours of Le Mans, driven by Graham Hill and Richie Ginther. It averaged 107.8 mph (173.5 km/h) and had a top speed of 142 mph (229 km/h). American Ray Heppenstall joined Howmet Corporation and McKee Engineering together to develop their own gas turbine sports car in 1968, the Howmet TX, which ran several American and European events, including two wins, and also participated in the 1968 24 Hours of Le Mans. The cars used Continental gas turbines, which eventually set six FIA land speed records for turbine-powered cars.

For open wheel racing, 1967's revolutionary STP-Paxton Turbocar fielded by racing and entrepreneurial legend Andy Granatelli and driven by Parnelli Jones nearly won the Indianapolis 500; the Pratt & Whitney ST6B-62 powered turbine car was almost a lap ahead of the second place car when a gearbox bearing failed just three laps from the finish line. The next year the STP Lotus 56 turbine car won the Indianapolis 500 pole position even though new rules restricted the air intake dramatically. In 1971, Lotus principal Colin Chapman introduced the Lotus 56B F1 car, powered by a Pratt & Whitney STN 6/76 gas turbine. Chapman had a reputation of building radical championship-winning cars, but had to abandon the project because there were too many problems with turbo lag.

Buses

The arrival of the Capstone Microturbine has led to several hybrid bus designs, starting with HEV-1 by AVS of Chattanooga, Tennessee in 1999, and closely followed by Ebus and ISE Research in California, and DesignLine Corporation in New Zealand (and later the United States). AVS turbine hybrids were plagued with reliability and quality control problems, resulting in liquidation of AVS in 2003. The most successful design by Designline is now operated in 5 cities in 6 countries, with over 30 buses in operation worldwide, and order for several hundred being delivered to Baltimore, and New York City.

Brescia Italy is using serial hybrid buses powered by microturbines on routes through the historical sections of the city.

Motorcycles

The MTT Turbine Superbike appeared in 2000 (hence the designation of Y2K Superbike by MTT) and is the first production motorcycle powered by a turbine engine - specifically, a Rolls-Royce Allison model 250 turboshaft engine, producing about 283 kW (380 bhp). Speed-tested to 365 km/h or 227 mph (according to some stories, the testing team ran out of road during the test), it holds the Guinness World Record for most powerful production motorcycle and most expensive production motorcycle, with a price tag of US$185,000.

Trains

Several locomotive classes have been powered by gas turbines, the most recent incarnation being Bombardier's JetTrain.

Tanks

The Third Reich *Wehrmacht Heer*'s development division, the Heereswaffenamt (Army Ordnance

Board), studied a number of gas turbine engine designs for use in tanks starting in mid-1944. The first gas turbine engine design intended for use in armored fighting vehicle propulsion, the BMW 003-based GT 101, was meant for installation in the Panther tank.

Marines from 1st Tank Battalion load a Honeywell AGT1500 multi-fuel turbine back into an M1 Abrams tank at Camp Coyote.

The second use of a gas turbine in an armored fighting vehicle was in 1954 when a unit, PU2979, specifically developed for tanks by C. A. Parsons & Co., was installed and trialled in a British Conqueror tank. The Stridsvagn 103 was developed in the 1950s and was the first mass-produced main battle tank to use a turbine engine. Since then, gas turbine engines have been used as APUs in some tanks and as main powerplants in Soviet/Russian T-80s and U.S. M1 Abrams tanks, among others. They are lighter and smaller than diesels at the same sustained power output but the models installed to date are less fuel efficient than the equivalent diesel, especially at idle, requiring more fuel to achieve the same combat range. Successive models of M1 have addressed this problem with battery packs or secondary generators to power the tank's systems while stationary, saving fuel by reducing the need to idle the main turbine. T-80s can mount three large external fuel drums to extend their range. Russia has stopped production of the T-80 in favor of the diesel-powered T-90 (based on the T-72), while Ukraine has developed the diesel-powered T-80UD and T-84 with nearly the power of the gas-turbine tank. The French Leclerc MBT's diesel powerplant features the "Hyperbar" hybrid supercharging system, where the engine's turbocharger is completely replaced with a small gas turbine which also works as an assisted diesel exhaust turbocharger, enabling engine RPM-independent boost level control and a higher peak boost pressure to be reached (than with ordinary turbochargers). This system allows a smaller displacement and lighter engine to be used as the tank's powerplant and effectively removes turbo lag. This special gas turbine/turbocharger can also work independently from the main engine as an ordinary APU.

A turbine is theoretically more reliable and easier to maintain than a piston engine since it has a simpler construction with fewer moving parts, but in practice, turbine parts experience a higher wear rate due to their higher working speeds. The turbine blades are highly sensitive to dust and fine sand so that in desert operations air filters have to be fitted and changed several times daily. An improperly fitted filter, or a bullet or shell fragment that punctures the filter, can damage the engine. Piston engines (especially if turbocharged) also need well-maintained filters, but they are more resilient if the filter does fail.

Like most modern diesel engines used in tanks, gas turbines are usually multi-fuel engines.

Marine Applications

Naval

The Gas turbine from MGB.

Gas turbines are used in many naval vessels, where they are valued for their high power-to-weight ratio and their ships' resulting acceleration and ability to get underway quickly.

The first gas-turbine-powered naval vessel was the Royal Navy's Motor Gun Boat *MGB 2009* (formerly *MGB 509*) converted in 1947. Metropolitan-Vickers fitted their F2/3 jet engine with a power turbine. The Steam Gun Boat *Grey Goose* was converted to Rolls-Royce gas turbines in 1952 and operated as such from 1953. The Bold class Fast Patrol Boats *Bold Pioneer* and *Bold Pathfinder* built in 1953 were the first ships created specifically for gas turbine propulsion.

The first large-scale, partially gas-turbine powered ships were the Royal Navy's Type 81 (Tribal class) frigates with combined steam and gas powerplants. The first, HMS *Ashanti* was commissioned in 1961.

The German Navy launched the first *Köln*-class frigate in 1961 with 2 Brown, Boveri & Cie gas turbines in the world's first combined diesel and gas propulsion system.

The Danish Navy had 6 *Søløven*-class torpedo boats (the export version of the British Brave class fast patrol boat) in service from 1965 to 1990, which had 3 Bristol Proteus (later RR Proteus) Marine Gas Turbines rated at 9,510 kW (12,750 shp) combined, plus two General Motors Diesel engines, rated at 340 kW (460 shp), for better fuel economy at slower speeds. And they also produced 10 Willemoes Class Torpedo / Guided Missile boats (in service from 1974 to 2000) which had 3 Rolls Royce Marine Proteus Gas Turbines also rated at 9,510 kW (12,750 shp), same as the Søløven-class boats, and 2 General Motors Diesel Engines, rated at 600 kW (800 shp), also for improved fuel economy at slow speeds.

The Swedish Navy produced 6 Spica-class torpedo boats between 1966 and 1967 powered by 3 Bristol Siddeley Proteus 1282 turbines, each delivering 3,210 kW (4,300 shp). They were later joined by 12 upgraded Norrköping class ships, still with the same engines. With their aft torpedo tubes replaced by antishipping missiles they served as missile boats until the last was retired in 2005.

The Finnish Navy commissioned two *Turunmaa*-class corvettes, *Turunmaa* and *Karjala*, in 1968. They were equipped with one 16,410 kW (22,000 shp) Rolls-Royce Olympus TM1 gas turbine and three Wärtsilä marine diesels for slower speeds. They were the fastest vessels in the Finnish Navy; they regularly achieved speeds of 35 knots, and 37.3 knots during sea trials. The *Turunmaa*s were decommissioned in 2002. *Karjala* is today a museum ship in Turku, and *Turunmaa* serves as a floating machine shop and training ship for Satakunta Polytechnical College.

The next series of major naval vessels were the four Canadian *Iroquois*-class helicopter carrying destroyers first commissioned in 1972. They used 2 ft-4 main propulsion engines, 2 ft-12 cruise engines and 3 Solar Saturn 750 kW generators.

An LM2500 gas turbine on USS Ford.

The first U.S. gas-turbine powered ship was the U.S. Coast Guard's *Point Thatcher*, a cutter commissioned in 1961 that was powered by two 750 kW (1,000 shp) turbines utilizing controllable-pitch propellers. The larger *Hamilton*-class High Endurance Cutters, was the first class of larger cutters to utilize gas turbines, the first of which (USCGC *Hamilton*) was commissioned in 1967. Since then, they have powered the U.S. Navy's *Oliver Hazard Perry*-class frigates, *Spruance* and *Arleigh Burke*-class destroyers, and *Ticonderoga*-class guided missile cruisers. USS *Makin Island*, a modified *Wasp*-class amphibious assault ship, is to be the Navy's first amphibious assault ship powered by gas turbines. The marine gas turbine operates in a more corrosive atmosphere due to the presence of sea salt in air and fuel and use of cheaper fuels.

Civilian Maritime

Up to the late 1940s, much of the progress on marine gas turbines all over the world took place in design offices and engine builder's workshops and development work was led by the British Royal Navy and other Navies. While interest in the gas turbine for marine purposes, both naval and mercantile, continued to increase, the lack of availability of the results of operating experience on early gas turbine projects limited the number of new ventures on seagoing commercial vessels being embarked upon. In 1951, the Diesel-electric oil tanker *Auris*, 12,290 Deadweight tonnage (DWT) was used to obtain operating experience with a main propulsion gas turbine under service conditions at sea and so became the first ocean-going merchant ship to be powered by a gas turbine. Built by Hawthorn Leslie at Hebburn-on-Tyne, UK, in accordance with plans and specifications drawn up by the Anglo-Saxon Petroleum Company and launched on the UK's Princess Elizabeth's 21st birthday in 1947, the ship was designed with an engine room layout that would allow for the experimental use of heavy fuel in one of its high-speed engines, as well as the future substitution of one of its diesel engines by a gas turbine. The *Auris* operated commercially as a tanker for three-and-a-half years with a diesel-electric propulsion unit as originally commissioned, but in 1951 one of its four 824 kW (1,105 bhp) diesel engines – which were known as "Faith", "Hope", "Charity" and "Prudence" - was replaced by the world's first marine gas turbine engine, a 890 kW (1,200 bhp) open-cycle gas turbo-alternator built by British Thompson-Houston Company in Rugby. Following successful sea trials off the Northumbrian coast, the *Auris* set sail from Hebburn-on-Tyne in October 1951 bound for Port Arthur in the US and then Curacao in the southern Caribbean returning to Avonmouth after 44 days at sea, successfully completing her historic trans-Atlantic crossing. During this time at sea the gas turbine burnt diesel fuel and operated without an involuntary stop or mechanical difficulty of any kind. She subsequently visited Swansea, Hull, Rotterdam, Oslo and Southampton covering a total of 13,211 nautical miles. The *Auris* then had all of its power plants replaced with a 3,910 kW (5,250 shp) directly coupled gas turbine to become the first civilian ship to operate solely on gas turbine power.

Despite the success of this early experimental voyage the gas turbine did not replace the diesel engine as the propulsion plant for large merchant ships. At constant cruising speeds the diesel engine simply had no peer in the vital area of fuel economy. The gas turbine did have more success in Royal Navy ships and the other naval fleets of the world where sudden and rapid changes of speed are required by warships in action.

The United States Maritime Commission were looking for options to update WWII Liberty ships, and heavy-duty gas turbines were one of those selected. In 1956, the *John Sergeant* was lengthened and equipped with a General Electric 4,900 kW (6,600 shp) HD gas turbine with exhaust-gas regeneration, reduction gearing and a variable-pitch propeller. It operated for 9,700 hours using residual fuel(Bunker C) for 7,000 hours. Fuel efficiency was on a par with steam propulsion at 0.318 kg/kW (0.523 lb/hp) per hour, and power output was higher than expected at 5,603 kW (7,514 shp) due to the ambient temperature of the North Sea route being lower than the design temperature of the gas turbine. This gave the ship a speed capability of 18 knots, up from 11 knots with the original power plant, and well in excess of the 15 knot targeted. The ship made its first transatlantic crossing with an average speed of 16.8 knots, in spite of some rough weather along the way. Suitable Bunker C fuel was only available at limited ports because the quality of the fuel was of a critical nature. The fuel oil also had to be treated on board to reduce contaminants and this was a labor-intensive process that was not suitable for automation at the time. Ultimately, the variable-pitch propeller, which was of a new and untested design, ended the trial, as three consecutive annual inspections revealed stress-cracking. This did not reflect poorly on the marine-propulsion gas-turbine concept though, and the trial was a success overall. The success of this trial opened the way for more development by GE on the use of HD gas turbines for marine use with heavy fuels. The *John Sergeant* was scrapped in 1972 at Portsmouth PA.

Boeing Jetfoil 929-100-007 Urzela of TurboJET.

Boeing launched its first passenger-carrying waterjet-propelled hydrofoil Boeing 929, in April 1974. Those ships were powered by two Allison 501-KF gas turbines.

Between 1971 and 1981, Seatrain Lines operated a scheduled container service between ports on the eastern seaboard of the United States and ports in northwest Europe across the North Atlantic with four container ships of 26,000 tonnes DWT. Those ships were powered by twin Pratt & Whitney gas turbines of the FT 4 series. The four ships in the class were named *Euroliner*, *Eurofreighter*, *Asialiner* and *Asiafreighter*. Following the dramatic Organization of the Petroleum Exporting Countries (OPEC) price increases of the mid-1970s, operations were constrained by rising fuel costs. Some modification of the engine systems on those ships was undertaken to permit the burning of a lower grade of fuel (i.e., marine diesel). Reduction of fuel costs was successful using a different untested fuel in a marine gas turbine but maintenance costs increased with the fuel change. After 1981 the ships were sold and refitted with, what at the time, was more economical diesel-fueled engines but the increased engine size reduced cargo space.

The first passenger ferry to use a gas turbine was the GTS *Finnjet*, built in 1977 and powered by two Pratt & Whitney FT 4C-1 DLF turbines, generating 55,000 kW (74,000 shp) and propelling the ship to a speed of 31 knots. However, the Finnjet also illustrated the shortcomings of gas turbine propulsion in commercial craft, as high fuel prices made operating her unprofitable. After four years of service, additional diesel engines were installed on the ship to reduce running costs during the off-season. The Finnjet was also the first ship with a Combined diesel-electric and gas propulsion. Another example of commercial use of gas turbines in a passenger ship is Stena Line's HSS class fastcraft ferries. HSS 1500-class *Stena Explorer*, *Stena Voyager* and *Stena Discovery* vessels use combined gas and gas setups of twin GE LM2500 plus GE LM1600 power for a total of 68,000 kW (91,000 shp). The slightly smaller HSS 900-class *Stena Carisma*, uses twin ABB–STAL GT35 turbines rated at 34,000 kW (46,000 shp) gross. The *Stena Discovery* was withdrawn from service in 2007, another victim of too high fuel costs.

In July 2000 the *Millennium* became the first cruise ship to be propelled by gas turbines, in a combined diesel and gas configuration. The liner RMS Queen Mary 2 uses a combined diesel and gas configuration.

In marine racing applications the 2010 C5000 Mystic catamaran Miss GEICO uses two Lycoming T-55 turbines for its power system.

Advances in Technology

Gas turbine technology has steadily advanced since its inception and continues to evolve. Development is actively producing both smaller gas turbines and more powerful and efficient engines. Aiding in these advances are computer-based design (specifically CFD and finite element analysis) and the development of advanced materials: Base materials with superior high-temperature strength (e.g., single-crystal superalloys that exhibit yield strength anomaly) or thermal barrier coatings that protect the structural material from ever-higher temperatures. These advances allow higher compression ratios and turbine inlet temperatures, more efficient combustion and better cooling of engine parts.

Computational Fluid Dynamics (CFD) has contributed to substantial improvements in the performance and efficiency of Gas Turbine engine components through enhanced understanding of the complex viscous flow and heat transfer phenomena involved. For this reason, CFD is one of the key computational tool used in Design & development of gas turbine engines.

The simple-cycle efficiencies of early gas turbines were practically doubled by incorporating inter-cooling, regeneration (or recuperation), and reheating. These improvements, of course, come at the expense of increased initial and operation costs, and they cannot be justified unless the decrease in fuel costs offsets the increase in other costs. The relatively low fuel prices, the general desire in the industry to minimize installation costs, and the tremendous increase in the simple-cycle efficiency to about 40 percent left little desire for opting for these modifications.

On the emissions side, the challenge is to increase turbine inlet temperatures while at the same time reducing peak flame temperature in order to achieve lower NO_x emissions and meet the latest emission regulations. In May 2011, Mitsubishi Heavy Industries achieved a turbine inlet temperature of 1,600 °C on a 320 megawatt gas turbine, and 460 MW in gas turbine combined-cycle power generation applications in which gross thermal efficiency exceeds 60%.

Compliant foil bearings were commercially introduced to gas turbines in the 1990s. These can withstand over a hundred thousand start/stop cycles and have eliminated the need for an oil system. The application of microelectronics and power switching technology have enabled the development of commercially viable electricity generation by microturbines for distribution and vehicle propulsion. The following are advantages and disadvantages of gas-turbine engines:

Advantages

- Very high power-to-weight ratio compared to reciprocating engines.

- Smaller than most reciprocating engines of the same power rating.

- Smooth rotation of the main shaft produces far less vibration than a reciprocating engine.

- Fewer moving parts than reciprocating engines results in lower maintenance cost and higher reliability/availability over its service life.

- Greater reliability, particularly in applications where sustained high power output is required.

- Waste heat is dissipated almost entirely in the exhaust. This results in a high-temperature exhaust stream that is very usable for boiling water in a combined cycle, or for cogeneration.

- Lower peak combustion pressures than reciprocating engines in general.

- High shaft speeds in smaller "free turbine units", although larger gas turbines employed in power generation operate at synchronous speeds.

- Low lubricating oil cost and consumption.

- Can run on a wide variety of fuels.

- Very low toxic emissions of CO and HC due to excess air, complete combustion and no "quench" of the flame on cold surfaces.

Disadvantages

- Core engine costs can be high due to use of exotic materials.

- Less efficient than reciprocating engines at idle speed.

- Longer startup than reciprocating engines.

- Less responsive to changes in power demand compared with reciprocating engines.

- Characteristic whine can be hard to suppress.

Testing

British, German, other national and international test codes are used to standardize the procedures and definitions used to test gas turbines. Selection of the test code to be used is an agreement between the purchaser and the manufacturer, and has some significance to the design of the turbine and associated systems. In the United States, ASME has produced several performance test codes on gas turbines. This includes ASME PTC 22-2014. These ASME performance test codes have gained international recognition and acceptance for testing gas turbines. The single most important and differentiating characteristic of ASME performance test codes, including PTC 22, is that the test uncertainty of the measurement indicates the quality of the test and is not to be used as a commercial tolerance.

Rotatory Engine

The rotary engine was an early type of internal combustion engine, usually designed with an odd number of cylinders per row in a radial configuration, in which the crankshaft remained stationary in operation, with the entire crankcase and its attached cylinders rotating around it as a unit. Its main application was in aviation, although it also saw use before its primary aviation role, in a few early motorcycles and automobiles.

An 80 horsepower (60 kW) rated Le Rhône 9C, a typical rotary engine of WWI.
The copper pipes carry the fuel-air mixture from the crankcase to the cylinder heads
acting collectively as an intake manifold.

This type of engine was widely used as an alternative to conventional inline engines (straight or V) during World War I and the years immediately preceding that conflict. It has been described as "a very efficient solution to the problems of power output, weight, and reliability".

By the early 1920s, the inherent limitations of this type of engine had rendered it obsolete.

This Le Rhône 9C installed on a Sopwith Pup fighter aircraft at the Fleet Air Arm Museum.
Note the narrowness of the mounting pedestal to the fixed crankshaft, and the size of the engine.

Distinction between "Rotary" and "Radial" Engines

A rotary engine is essentially a standard Otto cycle engine, with cylinders arranged radially around a central crankshaft just like a conventional radial engine, but instead of having a fixed cylinder block with rotating crankshaft as with a radial engine, the crankshaft remains stationary and the entire cylinder block rotates around it. In the most common form, the crankshaft was fixed solidly to the airframe, and the propeller was simply bolted to the front of the crankcase.

This difference also has much impact on design (lubrication, ignition, fuel admission, cooling, etc.) and functioning.

The Musée de l'Air et de l'Espace in Paris has on display a special, "sectioned" working model of an engine with seven radially disposed cylinders. It alternates between rotary and radial modes to demonstrate the difference between the internal motions of the two types of engine.

Arrangement

Like "fixed" radial engines, rotaries were generally built with an odd number of cylinders (usually 5, 7 or 9), so that a consistent every-other-piston firing order could be maintained, to provide smooth running. Rotary engines with an even number of cylinders were mostly of the "two row" type.

Most rotary engines were arranged with the cylinders pointing outwards from a single crankshaft, in the same general form as a radial, but there were also rotary boxer engines and even one-cylinder rotaries.

Advantages and Drawbacks

Three key factors contributed to the rotary engine's success at the time:

- Smooth running: Rotaries delivered power very smoothly because (relative to the engine mounting point) there are no reciprocating parts, and the relatively large rotating mass of the crankcase/cylinders (as a unit) acted as a flywheel.

- Improved cooling: when the engine was running, the rotating crankcase/cylinder assembly created its own fast-moving cooling airflow, even with the aircraft at rest.

- Weight advantage: many conventional engines had to have heavy flywheels added to smooth out power impulses and reduce vibration. Rotary engines gained a substantial power-to-weight ratio advantage by having no need for an added flywheel. They shared with other radial configuration engines the advantage of a small, flat crankcase, and because of their efficient air-cooling system, cylinders could be made with thinner walls and shallower cooling fins, which further reduced their weight.

Engine designers had always been aware of the many limitations of the rotary engine so when the static style engines became more reliable and gave better specific weights and fuel consumption, the days of the rotary engine were numbered.

- Rotary engines had a fundamentally inefficient total-loss oiling system. In order to reach the whole engine, the lubricating medium needed to enter the crankcase through the hollow crankshaft; but the centrifugal force of the revolving crankcase was directly opposed to any re-circulation. The only practical solution was for the lubricant to be aspirated with the fuel/air mixture, as in a two-stroke engine.

- Power increase also came with mass and size increases, multiplying gyroscopic precession from the rotating mass of the engine. This produced stability and control problems in aircraft in which these engines were installed, especially for inexperienced pilots.

- Power output increasingly went into overcoming the air-resistance of the spinning engine.

- Engine controls were tricky, and resulted in fuel waste.

The late WWI Bentley BR2, as the largest and most powerful rotary engine, had reached a point beyond which this type of engine could not be further developed, and it was the last of its kind to be adopted into RAF service.

Rotary Engine Control

Monosoupape Rotaries

It is often asserted that rotary engines had no throttle and hence power could only be reduced by intermittently cutting the ignition using a "blip" switch. This was almost literally true of the "Monosoupape" (single valve) type, which took most of the air into the cylinder through the exhaust valve, which remained open for a portion of the downstroke of the piston. Thus, the richness of the mixture in the cylinder could not be controlled via the crankcase intake. The "throttle" (fuel valve) of a monosoupape provided only a very limited degree of speed regulation, as opening it made the mixture too rich, while closing it made it too lean (in either case quickly stalling the engine, or damaging the cylinders). Early models featured a pioneering form of variable valve timing in an attempt to give greater control, but this caused the valves to burn and therefore it was abandoned.

The only way of running a Monosoupape engine smoothly at reduced revs was with a switch that changed the normal firing sequence so that each cylinder fired only once per two or three engine revolutions, but the engine remained more or less in balance. As with excessive use of the "blip" switch: running the engine on such a setting for too long resulted in large quantities of unburned fuel and oil in the exhaust, and gathering in the lower cowling, where it was a notorious fire hazard.

Normal Rotaries

Most rotaries had normal inlet valves, so that the fuel (and lubricating oil) was taken into the cylinders already mixed with air - as in a normal four-stroke engine. Although a conventional carburetor, with the ability to keep the fuel/air ratio constant over a range of throttle openings, was precluded by the spinning crankcase; it was possible to adjust the air supply through a separate flap valve or "bloctube". The pilot needed to set the throttle to the desired setting (usually full open) and then adjust the fuel/air mixture to suit using a separate "fine adjustment" lever that controlled the air supply valve (in the manner of a manual choke control). Due to the rotary engine's large rotational inertia, it was possible to adjust the appropriate fuel/air mixture by trial and error without stalling it, although this varied between different types of engine, and in any case it required a good deal of practice to acquire the necessary knack. After starting the engine with a known setting that allowed it to idle, the air valve was opened until maximum engine speed was obtained.

Throttling a running engine back to reduce revs was possible by closing off the fuel valve to the required position while re-adjusting the fuel/air mixture to suit. This process was also tricky, so that reducing power, especially when landing, was often accomplished instead by intermittently cutting the ignition using the blip switch.

Cutting cylinders using ignition switches had the drawback of letting fuel continue to pass through the engine, oiling up the spark plugs and making smooth restarting problematic. Also, the raw oil-fuel mix could collect in the cowling. As this could cause a serious fire when the switch was released, it became common practice for part or all of the bottom of the basically circular cowling on most rotary engines to be cut away, or fitted with drainage slots.

By 1918, a Clerget handbook advised maintaining all necessary control by using the fuel and air controls, and starting and stopping the engine by turning the fuel on and off. The recommended landing procedure involved shutting off the fuel using the fuel lever, while leaving the blip switch on. The windmilling propeller made the engine continue to spin without delivering any power as the aircraft descended. It was important to leave the ignition on to allow the spark plugs to continue to spark and keep them from oiling up, so that the engine could (if all went well) be restarted simply by re-opening the fuel valve. Pilots were advised to not use an ignition cut out switch, as it would eventually damage the engine.

Pilots of surviving or reproduction aircraft fitted with rotary engines still find that the blip switch is useful while landing, as it provides a more reliable, quicker way to initiate power if needed, rather than risk a sudden engine stall, or the failure of a windmilling engine to restart at the worst possible moment.

Use in Cars and Motorcycles

Although rotary engines were mostly used in aircraft, a few cars and motorcycles were built with rotary engines. Perhaps the first was the Millet motorcycle of 1892. A famous motorcycle, winning many races, was the Megola, which had a rotary engine inside the front wheel. Another motorcycle with a rotary engine was Charles Redrup's 1912 Redrup Radial, which was a three-cylinder 303 cc rotary engine fitted to a number of motorcycles by Redrup.

In 1904, the Barry engine, also designed by Redrup, was built in Wales: a rotating 2-cylinder boxer engine weighing 6.5 kg was mounted inside a motorcycle frame.

The early-1920s, German Megola motorcycle used a five-cylinder rotary engine within its front wheel design.

In the 1940s, Cyril Pullin developed the Powerwheel, a wheel with a rotating one-cylinder engine, clutch and drum brake inside the hub, but it never entered production.

Cars with rotary engines were built by American companies Adams-Farwell, Bailey, Balzer and Intrepid, amongst others.

Other Rotary Engines

Besides the configuration of cylinders moving around a fixed crankshaft, several different engine designs are also called *rotary engines*. The most notable pistonless rotary engine, the Wankel rotary engine has been used by NSU in the Ro80 car, by Mazda in a variety of cars such as the RX-series, and in some experimental aviation applications.

In the late 1970s, a concept engine called the Bricklin-Turner Rotary Vee was tested. The Rotary Vee is similar in configuration to the elbow steam engine. Piston pairs connect as solid V shaped members, with each end floating in a pair of rotating cylinders clusters. The rotating cylinder cluster pairs are set with their axes at a wide V angle. The pistons in each cylinder cluster move parallel to each other instead of a radial direction, This engine design has not gone into production. The Rotary Vee was intended to power the Bricklin SV-1.

Wankel Engine

Wankel engine is an Internal combustion engine unlike the piston cylinder arrangement. This engine uses the eccentric rotor design which directly converts the pressure energy of gases into rotatory motion. While in the piston-cylinder arrangement, the linear motion of the piston is used to convert into rotatory motion of crankshaft.

Basically, in a simple way, the rotor revolve in housings shaped in a fat figure-of-eight.

Parts of a Wankel Engine

- Rotor: The rotor has three convex faces which acts like a piston. The 3 corners of rotor forms a seal to the outside of the combustion chamber. It also has internal gear teeth in the centre on one side. This allows the rotor to revolve around a fix shaft.

- Housing: The housing is epitrochoidal in shape(roughly oval). The housing is cleverly designed as the 3 tips or corners of the rotor always stay in contact with the housing. The intake and exhaust ports are located in the housing.

- Inlet & exhaust ports: The intake port lets fresh mixture enter into combustion chamber & the exhaust gases expel out through outlet/exhaust port.

- Spark plug: A spark plug delivers electric current to the combustion chamber which ignites the air-fuel mixture leading to abrupt expansion of gas.

- Output shaft: The output shaft has eccentric lobes mounted on it, which means they are offset from centreline of the shaft. The rotor is not in pure rotation, but we need these eccentric lobes for pure rotation of the shaft.

Working

Intake

When a tip of the rotor passes the intake port, fresh mixture starts entering into the first chamber. The chamber draws fresh air until the second apex reaches the intake port & closes it. At the moment, fresh air-fuel mixture is sealed into first chamber & is being taken away for combustion.

Compression

The chamber one(between corner 1 to corner 2) containing the fresh charge gets compressed due to shape of the engine by the time it reaches to spark plug.

While this happens, a new mixture starts entering into the second chamber (between corner 2 to corner 3).

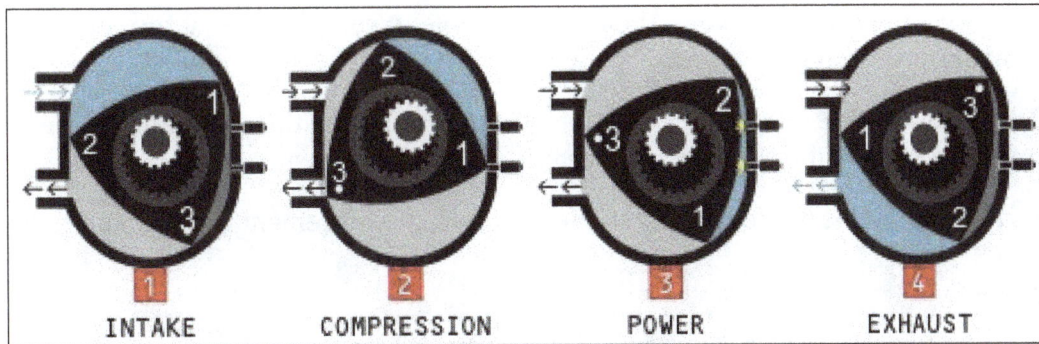

The four strokes of the engine with the corners numbered.

Combustion

When the spark plug ignites, the highly compressed mixture expands explosively. The pressure of expansion pushes the rotor in forward direction. This happens until the first corner passes through the exhaust port.

Exhaust

As the peak or corner 1 passes exhaust port, the hot high pressure combustion gases are free to flow out of the port. As the rotor continues to move, the volume of chamber goes on decreasing forcing the remaining gases out of port. By the time the corner 2 closes the exhaust port, corner 1 passes by the intake port repeating the cycle.

While the first chamber is discharging gases, the second chamber(between corner 2 to corner 3) is under compression. Simultaneously, chamber 3(between corner 3 to corner 1) is drawing fresh mixture.

This is the beauty of the engine – the four sequences of the four stroke cycle, which occur consecutively in a piston engine, occur simultaneously in the Wankel engine, producing power in a continuous stream.

Advantages

- Wankel engine has a very few moving parts; far less than 4 stroke piston engine. This makes the design of the engine simpler & the engine reliable.

- It is approximately 1/3rd of the size of the piston engines delivering same power output.

- Able to reach higher revolutions per minute than a piston engine.

- Wankel engine weighs almost 1/3rd of the weight of the piston engines delivering same power output. This leads to a higher power to weight ratio.

Disadvantages

- As each section has temperature differences, the material expansion of housing is different at different region. Therefore, the rotor is unable to completely seal the chamber in high temperature region sometimes.

- The combustion is slow as the combustion chamber is long, thin, and moving. Hence, there might be a possibility that the fresh charge discharges out without even burning.

- As unburnt fuel is in the exhaust stream, emissions requirements are difficult to meet.

References

- Berni kühne kuehne@tobe4u.de. "a new engine generation is born kottmann-motor-team six-stroke-engine. Accessed january 2008". Sechstaktmotor.de. Retrieved 2014-01-31

- Powerplants-reciprocating-engines: flightliteracy.com, Retrieved 31 March, 2019

- "A brilliant six-stroke from techies". 14 february 2007. Archived from the original on 22 february 2013. Retrieved 8 may2012

- How-does-a-wankel-engine-work: mechstuff.com, Retrieved 31 March, 2019

- Https://clubtechnical.com/advantages-and-disadvantages-of-reciprocating-internal-combustion-engines

- "Compression ratio theory in petrol and diesel engines explained with diagram". Crankit. 2014-04-03. Retrieved 2017-10-07

- Konrad reif (ed.): dieselmotor-management – systeme komponenten und regelung, 5th edition, springer, wiesbaden 2012, isbn 978-3-8348-1715-0, p. 286

- Jet-engine, technology: britannica.com, Retrieved 8 January, 2019

3

Components of Internal Combustion Engine

Some of the different components of an internal combustion engine are piston, supercharger, turbocharger, air filter, manifold absolute pressure sensor, starter, high tension leads and mass flow sensor. This chapter closely examines these components of internal combustion engine to provide an extensive understanding of the subject.

Main Parts of an Internal Combustion Engine

Cylinder Block

Cylinder is the main body of IC engine. Cylinder is a part in which the intake of fuel, compression of fuel and burning of fuel take place. The main function of cylinder is to guide the piston. It is in direct contact with the products of combustion so it must be cooled. For cooling of cylinder a water jacket (for liquid cooling used in most of cars) or fin (for air cooling used in most of bikes) are situated at the outer side of cylinder. At the upper end of cylinder, cylinder head and at the bottom end crank case is bolted. The upper side of cylinder is consists a combustion chamber where fuel burns. To handle all this pressure and temperature generated by combustion of fuel, cylinder material should have high compressive strength. So it is made by high grade cast iron. It is made by casting and usually cast in one piece.

Cylinder Head

The top end of the engine cylinder is closed by means of removable cylinder head. There are two holes or ports at the cylinder head, one for intake of fuel and other for exhaust. Both the intake and exhaust ports are closed by the two valves known as inlet and exhaust valve. The inlet valve, exhaust valve, spark plug, injector etc. are bolted on the cylinder head. The main function of cylinder head is to seal the cylinder block and not to permit entry and exit of gases on cover head valve engine. Cylinder head is usually made by cast iron or aluminum. It is made by casting or forging and usually in one piece.

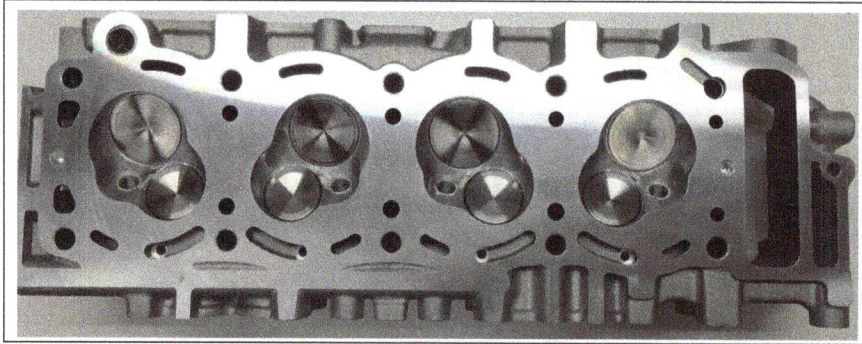

Connecting Rod

Connecting rod connects the piston to crankshaft and transmits the motion and thrust of piston to crankshaft. It converts the reciprocating motion of the piston into rotary motion of crankshaft. There are two end of connecting rod; one is known as big end and other as small end. Big end is connected to the crankshaft and the small end is connected to the piston by use of piston pin. The connecting rods are made of nickel, chrome, and chrome vanadium steels. For small engines the material may be aluminum.

Crankshaft

The crankshaft of an internal combustion engine receives the efforts or thrust supplied by piston to the connecting rod and converts the reciprocating motion of piston into rotary motion of crankshaft. The crankshaft mounts in bearing so it can rotate freely. The shape and size of crankshaft depends on the number and arrangement of cylinders. It is usually made by steel forging, but some makers use special types of cast-iron such as spheroidal graphitic or nickel alloy castings which are cheaper to produce and have good service life.

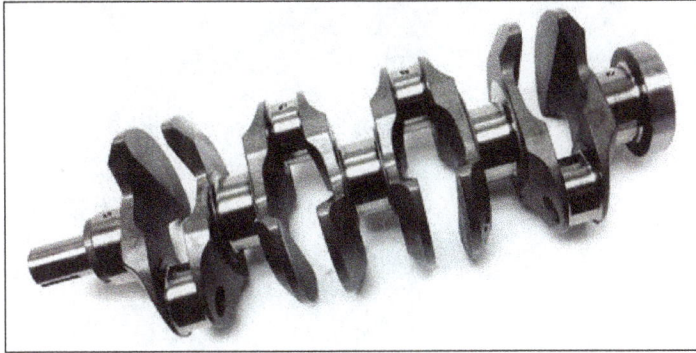

Engine Bearing

Everywhere there is rotary action in the engine, bearings are needed. Bearings are used to support the moving parts. The crankshaft is supported by bearing. The connecting rod big end is attached to the crank pin on the crank of the crankshaft by a bearing. A piston pin at the small end is used to attach the rod to the piston is also rides in bearings. The main function of bearings is to reduce friction between these moving parts. In an IC engine sliding and rolling types of bearing used. The sliding type bearing which are sometime called bush is use to attach the connecting rod to the piston and crankshaft. They are split in order to permit their assembly into the engine. The rolling and ball bearing is used to support crankshaft so it can rotate freely. The typical bearing half is made of steel or bronze back to which a lining of relatively soft bearing material is applied.

Crankcase

The main body of the engine at which the cylinder are attached and which contains the crankshaft and crankshaft bearing is called crankcase. It serves as the lubricating system too and sometime it is called oil sump. All the oil for lubrication is placed in it.

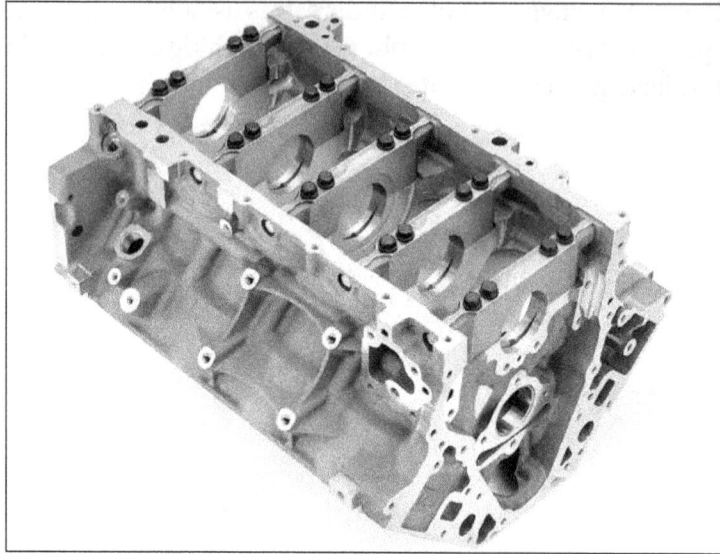

Valves

To control the inlet and exhaust of internal combustion engine, valves are used. The number of valves in an engine depends on the number of cylinders. Two valves are used for each cylinder one for inlet of air-fuel mixture inside the cylinder and other for exhaust of combustion gases. The valves are fitted in the port at the cylinder head by use of strong spring. This spring keep them closed. Both valves usually open inwards.

All types of valves.

Spark Plug

It is used in spark ignition engine. The main function of a spark plug is to conduct a high potential from the ignition system into the combustion chamber to ignite the compressed air fuel mixture. It is fitted on cylinder head. The spark plug consists of a metal shell having two electrodes which are insulated from each other with an air gap. When high potential current supply to spark plug it jumping from the supply electrode and produces the necessary spark.

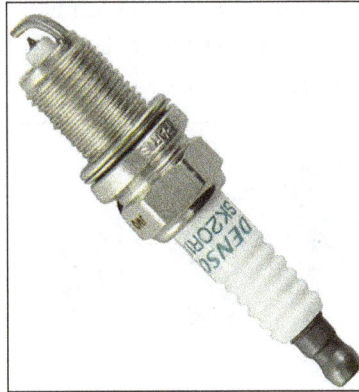

Injector

Injector is usually used in compression ignition engine. It sprays the fuel into combustion chamber at the end of compression stroke. It is fitted on cylinder head.

Manifold

The main function of manifold is to supply the air fuel mixture and collects the exhaust gases equally from all cylinder. In an internal combustion engine two manifold are used, one for intake and other for exhaust. They are usually made by aluminum alloy.

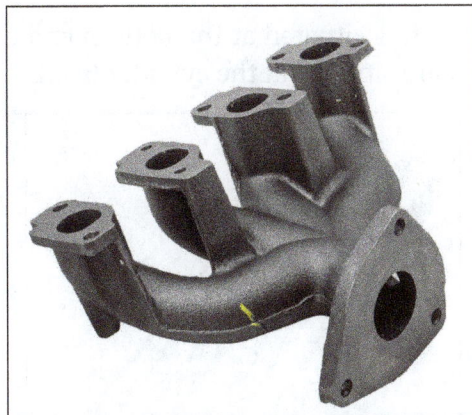

Camshaft

Camshaft is used in IC engine to control the opening and closing of valves at proper timing. For proper engine output inlet valve should open at the end of exhaust stroke and closed at the end of intake stroke. So to regulate its timing, a cam is use which is oval in shape and it exerts a pressure on the valve to open and release to close. It is drive by the timing belt which drives by crankshaft. It is placed at the top or at the bottom of cylinder.

Gudgeon Pin or Piston Pin

These are hardened steel parallel spindles fitted through the piston bosses and the small end bushes or eyes to allow the connecting rods to swivel. It connects the piston to connecting rod. It is made hollow for lightness.

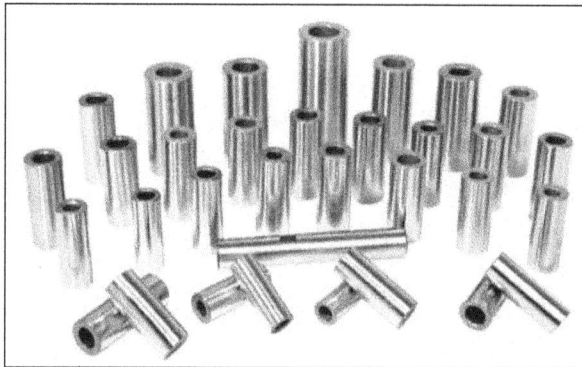

Pushrod

Pushrod is used when the camshaft is situated at the bottom end of cylinder. It carries the camshaft motion to the valves which are situated at the cylinder head.

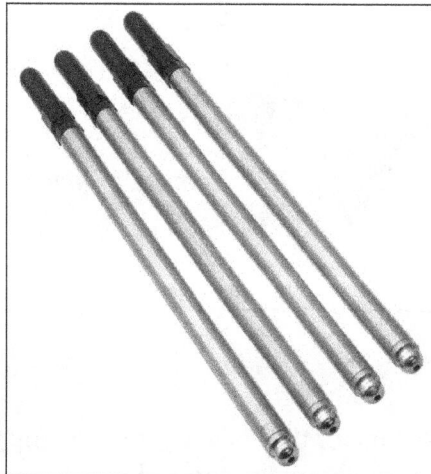

Flywheel

A flywheel is secured on the crankshaft. The main function of flywheel is to rotate the shaft during preparatory stroke. It also makes crankshaft rotation more uniform.

Piston

A piston is a component of reciprocating engines, reciprocating pumps, gas compressors and pneumatic cylinders, among other similar mechanisms. It is the moving component that is contained by a cylinder and is made gas-tight by piston rings. In an engine, its purpose is to transfer force from expanding gas in the cylinder to the crankshaft via a piston rod and/or connecting rod. In a pump, the function is reversed and force is transferred from the crankshaft to the piston for the purpose of compressing or ejecting the fluid in the cylinder. In some engines, the piston also acts as a valve by covering and uncovering ports in the cylinder.

Pistons within a sectioned petrol engine.

Piston Engines

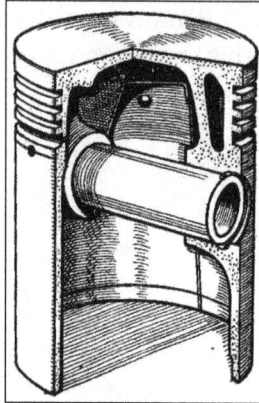

Internal combustion engine piston, sectioned to show the gudgeon pin.

An internal combustion engine is acted upon by the pressure of the expanding combustion gases in the combustion chamber space at the top of the cylinder. This force then acts downwards through the connecting rod and onto the crankshaft. The connecting rod is attached to the piston by a swivelling gudgeon pin (US: wrist pin). This pin is mounted within the piston: unlike the steam engine, there is no piston rod or crosshead (except big two stroke engines).

The typical piston design is on the picture. This type of piston is widely used in car diesel engines. According to purpose, supercharging level and working conditions of engines the shape and proportions can be changed.

High-power diesel engines work in difficult conditions. Maximum pressure in the combustion chamber can reach 20 MPa and maximum temperature of some piston surfaces can exceed 450°C. It is possible to improve piston cooling by creating special cooling cavity. Injector supplies this cooling cavity "A" with oil through oil supply channel "B". For better temperature reduction construction should be carefully calculated and analyzed. Oil flow in cooling cavity should be not less than 80% from oil flow through the injector.

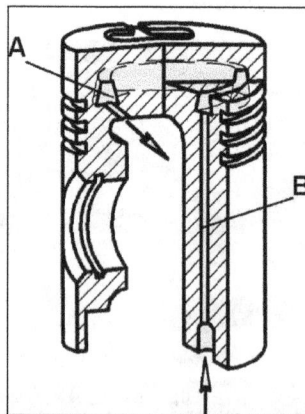

A: cooling cavity; B: oil supply channel.

The pin itself is of hardened steel and is fixed in the piston, but free to move in the connecting rod. A few designs use a 'fully floating' design that is loose in both components. All pins must be prevented from moving sideways and the ends of the pin digging into the cylinder wall, usually by circlips.

Gas sealing is achieved by the use of piston rings. These are a number of narrow iron rings, fitted loosely into grooves in the piston, just below the crown. The rings are split at a point in the rim, allowing them to press against the cylinder with a light spring pressure. Two types of ring are used: the upper rings have solid faces and provide gas sealing; lower rings have narrow edges and a U-shaped profile, to act as oil scrapers. There are many proprietary and detail design features associated with piston rings.

Pistons are cast from aluminium alloys. For better strength and fatigue life, some racing pistons may be forged instead. Billet pistons are also used in racing engines because they do not rely on the size and architecture of available forgings, allowing for last-minute design changes. Although not commonly visible to the naked eye, pistons themselves are designed with a certain level of ovality and profile taper, meaning they are not perfectly round, and their diameter is larger near the bottom of the skirt than at the crown.

Early pistons were of cast iron, but there were obvious benefits for engine balancing if a lighter alloy could be used. To produce pistons that could survive engine combustion temperatures, it was necessary to develop new alloys such as Y alloy and Hiduminium, specifically for use as pistons.

A few early gas engines had double-acting cylinders, but otherwise effectively all internal combustion engine pistons are single-acting. During World War II, the US submarine *Pompano* was fitted with a prototype of the infamously unreliable H.O.R. double-acting two-stroke diesel engine. Although compact, for use in a cramped submarine, this design of engine was not repeated.

Trunk Pistons

Trunk pistons are long relative to their diameter. They act both as a piston and cylindrical crosshead. As the connecting rod is angled for much of its rotation, there is also a side force that reacts along the side of the piston against the cylinder wall. A longer piston helps to support this.

Trunk pistons have been a common design of piston since the early days of the reciprocating internal combustion engine. They were used for both petrol and diesel engines, although high speed engines have now adopted the lighter weight slipper piston.

A characteristic of most trunk pistons, particularly for diesel engines, is that they have a groove for an oil ring below the gudgeon pin, in addition to the rings between the gudgeon pin and crown.

The name 'trunk piston' derives from the 'trunk engine', an early design of marine steam engine. To make these more compact, they avoided the steam engine's usual piston rod with separate crosshead and were instead the first engine design to place the gudgeon pin directly within the piston. Otherwise, these trunk engine pistons bore little resemblance to the trunk piston; they were extremely large diameter and double-acting. Their 'trunk' was a narrow cylinder mounted in the centre of the piston.

Crosshead Pistons

Large slow-speed Diesel engines may require additional support for the side forces on the piston. These engines typically use crosshead pistons. The main piston has a large piston rod extending downwards from the piston to what is effectively a second smaller-diameter piston. The main piston is responsible for gas sealing and carries the piston rings. The smaller piston is purely a mechanical guide. It runs within a small cylinder as a trunk guide and also carries the gudgeon pin.

Lubrication of the crosshead has advantages over the trunk piston as its lubricating oil is not subject to the heat of combustion: the oil is not contaminated by combustion soot particles, it does not break down owing to the heat and a thinner, less viscous oil may be used. The friction of both piston and crosshead may be only half of that for a trunk piston.

Because of the additional weight of these pistons, they are not used for high-speed engines.

Slipper Pistons

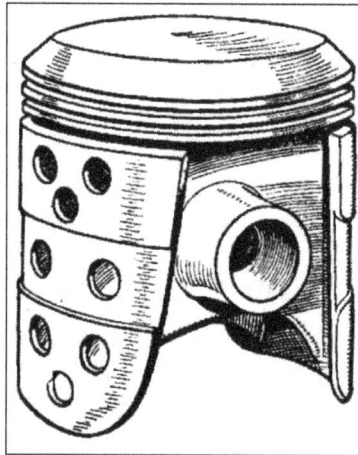

Slipper piston.

A slipper piston is a piston for a petrol engine that has been reduced in size and weight as much as possible. In the extreme case, they are reduced to the piston crown, support for the piston rings, and just enough of the piston skirt remaining to leave two lands so as to stop the piston rocking in the bore. The sides of the piston skirt around the gudgeon pin are reduced away from the cylinder wall. The purpose is mostly to reduce the reciprocating mass, thus making it easier to balance the engine and so permit high speeds. In racing applications, slipper piston skirts can be configured to yield extremely light weight while maintaining the rigidity and strength of a full skirt. Reduced inertia also improves mechanical efficiency of the engine: the forces required to accelerate and decelerate the reciprocating parts cause more piston friction with the cylinder wall than the fluid pressure on the piston head. A secondary benefit may be some reduction in friction with the cylinder wall, since the area of the skirt, which slides up and down in the cylinder is reduced by half. However, most friction is due to the piston rings, which are the parts which actually fit the tightest in the bore and the bearing surfaces of the wrist pin, and thus the benefit is reduced.

Deflector Pistons

Deflector pistons are used in two-stroke engines with crankcase compression, where the gas flow within the cylinder must be carefully directed in order to provide efficient scavenging. With cross scavenging, the transfer (inlet to the cylinder) and exhaust ports are on directly facing sides of the cylinder wall. To prevent the incoming mixture passing straight across from one port to the other, the piston has a raised rib on its crown. This is intended to deflect the incoming mixture upwards, around the combustion chamber.

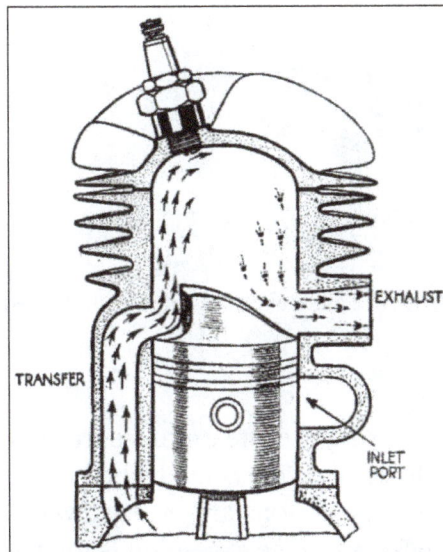

Two-stroke deflector piston.

Much effort, and many different designs of piston crown, went into developing improved scavenging. The crowns developed from a simple rib to a large asymmetric bulge, usually with a steep face on the inlet side and a gentle curve on the exhaust. Despite this, cross scavenging was never as effective as hoped. Most engines today use Schnuerle porting instead. This places a pair of transfer ports in the sides of the cylinder and encourages gas flow to rotate around a vertical axis, rather than a horizontal axis.

Racing Pistons

Early (c. 1830) piston for a beam engine. The piston seal is made by turns of wrapped rope.

In racing engines, piston strength and stiffness is typically much higher than that of a passenger car engine, while the weight is much less, to achieve the high engine RPM necessary in racing.

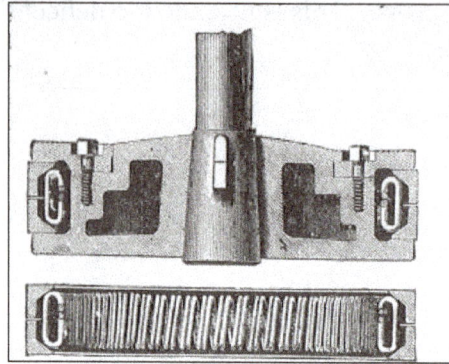

Cast-iron steam engine piston, with a metal piston ring spring-loaded against the cylinder wall.

Steam Engines

Steam engines are usually double-acting (i.e. steam pressure acts alternately on each side of the piston) and the admission and release of steam is controlled by slide valves, piston valves or poppet valves. Consequently, steam engine pistons are nearly always comparatively thin discs: their diameter is several times their thickness. (One exception is the trunk engine piston, shaped more like those in a modern internal-combustion engine.) Another factor is that since almost all steam engines use crossheads to translate the force to the drive rod, there are few lateral forces acting to try and "rock" the piston, so a cylinder-shaped piston skirt isn't necessary.

Supercharger

A supercharger is an air compressor that increases the pressure or density of air supplied to an internal combustion engine. This gives each intake cycle of the engine more oxygen, letting it burn more fuel and do more work, thus increasing power.

Power for the supercharger can be provided mechanically by means of a belt, gear, shaft, or chain connected to the engine's crankshaft.

Common usage restricts the term *supercharger* to mechanically driven units; when power is instead provided by a turbine powered by exhaust gas, a supercharger is known as a *turbocharger* or just a *turbo* - or in the past a *turbosupercharger*.

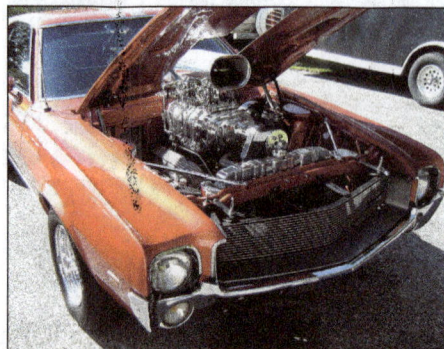

Roots type supercharger on AMC V8 engine for dragstrip racing.

Types of Supercharger

There are two main types of superchargers defined according to the method of gas transfer: positive displacement and dynamic compressors. Positive displacement blowers and compressors deliver an almost constant level of pressure increase at all engine speeds (RPM). Dynamic compressors do not deliver pressure at low speeds; above a threshold speed pressure increases exponentially.

Positive Displacement

An Eaton M62 Roots-type supercharger is visible at the front of this Ecotec LSJ engine in a 2006 Saturn Ion Red Line.

Lysholm screw rotors with complex shape of each rotor, which must run at high speed and with close tolerances. This makes this type of supercharger expensive.

Positive-displacement pumps deliver a nearly fixed volume of air per revolution at all speeds (minus leakage, which is almost constant at all speeds for a given pressure, thus its importance decreases at higher speeds).

Major types of positive-displacement pumps include:

- Roots

- Lysholm twin-screw

- Sliding vane

- Scroll-type supercharger, also known as the G-Lader

Compression Type

Positive-displacement pumps are further divided into internal and external compression types:

Roots superchargers, including high helix roots superchargers, produce compression externally.

- External compression refers to pumps that transfer air at ambient pressure. If an engine equipped with a supercharger that compresses externally is running under boost conditions, the pressure inside the supercharger remains at ambient pressure; air is only pressurized downstream of the supercharger. Roots superchargers tend to be very mechanically efficient at moving air at low-pressure differentials, whereas at high-pressure ratios, internal compression superchargers tend to be more mechanically efficient.

All the other types have some degree of internal compression.

- Internal compression refers to the compression of air within the supercharger itself, which, already at or close to boost level, can be delivered smoothly to the engine with little or no backflow. Internal compression devices usually use a fixed internal compression ratio. When the boost pressure is equal to the compression pressure of the supercharger, the backflow is zero. If the boost pressure exceeds that compression pressure, backflow can still occur as in a roots blower. The internal compression ratio of this type of supercharger can be matched to the expected boost pressure in order to optimize mechanical efficiency.

Capacity Rating

Positive-displacement superchargers are usually rated by their capacity per revolution. In the case of the Roots blower, the GMC rating pattern is typical. The GMC types are rated according to how many two-stroke cylinders, and the size of those cylinders, it is designed to scavenge. GMC has made 2–71, 3–71, 4–71, and the famed 6–71 blowers. For example, a 6–71 blower is designed to scavenge six cylinders of 71 cubic inches (1,163 cc) each and would be used on a two-stroke diesel of 426 cubic inches (6,981 cc), which is designated a 6–71; the blower takes this same designation. However, because 6–71 is actually the *engine's* designation, the actual displacement is less than the simple multiplication would suggest. A 6–71 actually pumps 339 cubic inches (5,555 cc) per revolution (but as it spins faster than the engine, it can easily put out the same displacement as the engine per engine rev).

Aftermarket derivatives continue the trend with 8–71 to current 16–71 blowers used in different motorsports. From this, one can see that a 6–71 is roughly twice the size of a 3–71. GMC also made 53 cu in (869 cc) series in 2–, 3–, 4–, 6–, and 8–53 sizes, as well as a "V71" series for use on engines using a V configuration.

Dynamic

Dynamic compressors rely on accelerating the air to high speed and then exchanging that velocity for pressure by diffusing or slowing it down.

Major types of dynamic compressor are:

- Centrifugal

- Variable ratio centrifugal

- Multi-stage axial-flow

- Pressure wave supercharger

Supercharger Drive Types

Superchargers are further defined according to their method of drive.

- Belt (V-belt, Synchronous belt, Flat belt)

- Direct drive

- Gear drive

- Chain drive

Temperature Effects and Intercoolers

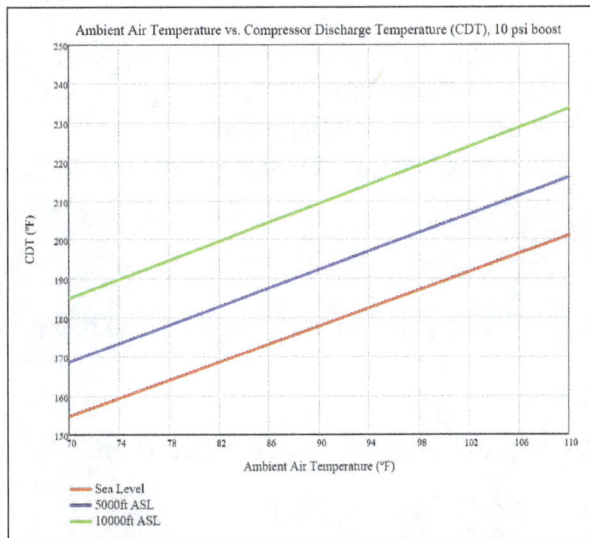

Supercharger CDT vs. ambient temperature. Graph shows how a supercharger's CDT varies with air temperature and altitude (absolute pressure).

One disadvantage of supercharging is that compressing the air increases its temperature. When a supercharger is used on an internal combustion engine, the temperature of the fuel/air charge becomes a major limiting factor in engine performance. Extreme temperatures will cause detonation of the fuel-air mixture (spark ignition engines) and damage to the engine. In cars, this can cause a problem when it is a hot day outside, or when an excessive level of boost is reached.

It is possible to estimate the temperature rise across a supercharger by modeling it as an isentropic process.

$$\frac{T_2}{T_1} = \left(\frac{p_2}{p_1} \right)^{\frac{\gamma - 1}{\gamma}}$$

where:

T_1 = ambient air temperature (absolute).

T_2 = temperature after the compressor (absolute).

p_1 = ambient atmospheric pressure (absolute).

p_2 = pressure after the compressor (absolute).

γ = Ratio of specific heat capacities = C_p / C_v = 1.4 for air.

C_p = Specific heat at constant pressure.

C_v = Specific heat at constant volume.

For example, if a supercharged engine is pushing 10 psi (0.69 bar) of boost at sea level (ambient pressure of 14.7 psi (1.01 bar), ambient temperature of 75 °F (24 °C)), the temperature of the air after the supercharger will be 160.5 °F (71.4 °C). This temperature is known as the compressor discharge temperature (CDT) and highlights why a method for cooling the air after the compressor is so important.

In the example above, the ambient air pressure (1.01bar) is added to the boost (0.69 bar) to get total pressure (1.70 bar), which is the value used for in the equation. The temperatures must be in absolute values, using the Kelvin scale, which begins at absolute zero (0 Kelvin) and where 0 °C is 273.15 K. A Kelvin unit is the same size as a Celsius degree (so 24 °C added to absolute zero is simply 273.15 K + 24 K).

So this means,

p_2 = 1.70 bar (24.7 psi = [14.7 psi + 10 psi boost]; or 1.70 bar = [1.01 bar + 0.69 bar]).

p_1 = 1.01 bar.

T_1 = 297.15K (24 K + 273.15 K; use the Kelvin scale, where 0 °C equals 273.15 Kelvin).

The exponent becomes 0.286 (or 1.4-1/[1.4]).

Resulting in:

T_2 = 344. 81 K, which is roughly 71.7 °C [344.81 K - 273.15 (since 273.15 K is 0 °C)]

Where, 71.7 °C exceeds 160 °F.

While it is true that higher intake temperatures for internal combustion engines will ingest air of lower density, this only holds correct for static, unchanging air pressure. i.e. on a hot day, an engine will intake less oxygen per engine cycle than it would on a cold day. However, the heating of the air, while in the supercharger compressor, does not reduce the density of the air due to its rise in temperature. The rise in temperature is due to its rise in pressure. Energy is being added to the air and this is seen in both its energy, internal to the molecules (temperature) and of the air in static pressure, as well as the velocity of the gas.

Inter-cooling makes no change in the density of the air after it has been compressed. It is only removing the thermal energy of the air from the compression process. i.e. the inter-cooler only removes the energy put in by the compression process and does not alter the density of air, so that the air/fuel mixture is not so hot that it causes it to ignite before the spark ignites it, otherwise known as pre-ignition.

Two-stroke Engines

In two-stroke engines, scavenging is required to purge exhaust gasses, as well as charge the cylinders for the next power stroke. In small engines this requirement is commonly met by using the crankcase as a blower; the descending piston during the power stroke compresses air in the crankcase used to purge the cylinder. Scavenging blowing should not be confused with supercharging, as no charge compression takes place. As the volume change produced by the lower side of the piston is the same as the upper face, this is limited to scavenging and cannot provide any supercharging.

Larger engines usually use a separate blower for scavenging and it was for this type of operation that the Roots blower has been utilized. Historically, many designs of blower have been used, from separate pumping cylinders, 'top hat' pistons combining two pistons of different diameter the larger one being used for scavenging, various rotary blowers, and centrifugal turbo-compressors, including turbochargers. Turbocharging two-stroke engines is difficult, but not impossible, as a turbocharger does not provide any boost until it has had time to spin up to speed. Purely turbocharged two-stroke engines may thus have difficulty when starting, with poor combustion and dirty exhausts, possibly even four-stroking. Some two-stroke turbochargers, notably those used on Electro-Motive Diesel locomotive engines, are mechanically driven at lower engine speeds through an overrunning clutch to provide adequate scavenging air. As engine speed and exhaust gas volume increase, the turbocharger no longer is dependent on mechanical drive and the overrunning clutch disengages.

Simple two-stroke engines with ported inlet and exhaust cannot be supercharged since the inlet port always closes first. For this reason, two-stroke Diesel engines usually have mechanical exhaust valves with separate timing to allow supercharging. Regardless of this, two-stroke engines require scavenging at all engine speeds and so turbocharged two-stroke engines must still employ a blower, usually Roots type. This blower may be mechanically or electrically driven, in either case, the blower may be disengaged once the turbocharger starts to deliver air.

Automobiles

1929 "Blower" Bentley. The large "blower" (supercharger),
located in front of the radiator, gave the car its name.

In 1900, Gottlieb Daimler, of Daimler-Benz (Daimler AG), was the first to patent a forced-induction system for internal combustion engines, superchargers based on the twin-rotor air-pump design, first patented by the American Francis Marion Roots in 1860, *the* basic design for the modern Roots type supercharger.

The first supercharged cars were introduced at the 1921 Berlin Motor Show: the 6/20 hp and 10/35 hp Mercedes. These cars went into production in 1923 as the 6/25/40 hp (regarded as the first supercharged road car) and 10/40/65 hp. These were normal road cars as other supercharged cars at same time were almost all racing cars, including the 1923 Fiat 805-405, 1923 Miller 122 1924 Alfa Romeo P2, 1924 Sunbeam, 1925 Delage, and the 1926 Bugatti Type 35C. At the end of the 1920s, Bentley made a supercharged version of the Bentley 4½ Litre road car. Since then, superchargers (and turbochargers) have been widely applied to racing and production cars, although the supercharger's technological complexity and cost have largely limited it to expensive, high-performance cars.

Supercharging versus Turbocharging

A G-Lader scroll-type supercharger on a Volkswagen Golf Mk1.

Keeping the air that enters the engine cool is an important part of the design of both superchargers and turbochargers. Compressing air increases its temperature, so it is common to use a small radiator called an intercooler between the pump and the engine to reduce the temperature of the air.

There are three main categories of superchargers for automotive use:

- Centrifugal turbochargers: Driven from exhaust gases.

- Centrifugal superchargers: Driven directly by the engine via a belt-drive.

- Positive displacement pumps: Such as the Roots, twin-screw (Lysholm), and TVS (Eaton) blowers.

Roots blowers tend to be only 40–50% efficient at high boost levels; by contrast centrifugal (dynamic) superchargers are 70–85% efficient at high boost. Lysholm-style blowers can be nearly as efficient as their centrifugal counterparts over a narrow range of load/speed/boost, for which the system must be specifically designed.

Mechanically driven superchargers may absorb as much as a third of the total crankshaft power of the engine and are less efficient than turbochargers. However, in applications for which engine response and power are more important than other considerations, such as top-fuel dragsters and vehicles used in tractor pulling competitions, mechanically driven superchargers are very common.

The thermal efficiency, or fraction of the fuel/air energy that is converted to output power, is less with a mechanically driven supercharger than with a turbocharger, because turbochargers use energy from the exhaust gas that would normally be wasted. For this reason, both economy and the power of a turbocharged engine are usually better than with superchargers.

Turbochargers suffer (to a greater or lesser extent) from so-called *turbo-spool* (turbo lag; more correctly, boost lag), in which initial acceleration from low RPM is limited by the lack of sufficient exhaust gas mass flow (pressure). Once engine RPM is sufficient to raise the turbine RPM into its designed operating range, there is a rapid increase in power, as higher turbo boost causes more exhaust gas production, which spins the turbo yet faster, leading to a belated "surge" of acceleration. This makes the maintenance of smoothly increasing RPM far harder with turbochargers than with engine-driven superchargers, which apply boost in direct proportion to the engine RPM. The main advantage of an engine with a mechanically driven supercharger is better throttle response, as well as the ability to reach full-boost pressure instantaneously. With the latest turbocharging technology and direct gasoline injection, throttle response on turbocharged cars is nearly as good as with mechanically powered superchargers, but the existing lag time is still considered a major drawback, especially considering that the vast majority of mechanically driven superchargers are now driven off clutched pulleys, much like an air compressor.

Turbocharging has been more popular than superchargers among auto manufacturers owing to better power and efficiency. For instance Mercedes-Benz and Mercedes-AMG previously had supercharged "Kompressor" offerings in the early 2000s such as the C230K, C32 AMG, and S55 AMG, but they have abandoned that technology in favor of turbocharged engines released around 2010 such as the C250 and S65 AMG biturbo. However, Audi did introduce its 3.0 TFSI supercharged V6 in 2009 for its A6, S4, and Q7, while Jaguar has its supercharged V8 engine available as a performance option in the XJ, XF, XKR, and F-Type, and, via joint ownership by Tata motors in the Range Rover also.

Twincharging

In the 1985 and 1986 World Rally Championships, Lancia ran the Delta S4, which incorporated both a belt-driven supercharger and exhaust-driven turbocharger. The design used a complex series of bypass valves in the induction and exhaust systems as well as an electromagnetic clutch so that, at low engine speeds, boost was derived from the supercharger. In the middle of the rev range, boost was derived from both systems, while at the highest revs the system disconnected the drive from the supercharger and isolated the associated ducting. This was done in an attempt to exploit the advantages of each of the charging systems while removing the disadvantages. In turn, this approach brought greater complexity and impacted on the car's reliability in WRC events, as well as increasing the weight of engine ancillaries in the finished design.

The Volkswagen TSI engine (or Twincharger) is a 1.4-litre direct-injection motor that also uses both a supercharger and turbocharger. Volvo offers a 2.0-liter engine with supercharger and turbocharged in hybrid models like S60, XC60 and XC90.

Aircraft

Altitude Effects

The Rolls-Royce Merlin, a supercharged aircraft engine from World War II.
The supercharger is at the rear of the engine at right.

Superchargers are a natural addition to aircraft piston engines that are intended for operation at high altitudes. As an aircraft climbs to a higher altitude, air pressure and air density decreases. The output of a piston engine drops because of the reduction in the mass of air that can be drawn into the engine. For example, the air density at 30,000 ft (9,100 m) is $\frac{1}{3}$ of that at sea level, thus only $\frac{1}{3}$ of the amount of air can be drawn into the cylinder, with enough oxygen to provide efficient combustion for only a third as much fuel. So, at 30,000 ft (9,100 m), only $\frac{1}{3}$ of the fuel burnt at sea level can be burnt.An advantage of the decreased air density is that the airframe experiences only about 1/3 of the aerodynamic drag. In addition, there is decreased back pressure on the exhaust gases. On the other hand, more energy is consumed holding an airplane up with less air in which to generate lift.

A supercharger can be thought of either as artificially increasing the density of the air by compressing it or as forcing more air than normal into the cylinder every time the piston moves down.

A Centrifugal supercharger of a Bristol Centaurus radial aircraft engine.

A supercharger compresses the air back to sea-level-equivalent pressures, or even much higher, in order to make the engine produce just as much power at cruise altitude as it does at sea level. With the reduced aerodynamic drag at high altitude and the engine still producing rated power, a supercharged airplane can fly much faster at altitude than a naturally aspirated one. The pilot controls the output of the supercharger with the throttle and indirectly via the propeller governor control. Since the size of the supercharger is chosen to produce a given amount of pressure at high altitude, the supercharger is oversized for low altitude. The pilot must be careful with the throttle and watch the manifold pressure gauge to avoid over-boosting at low altitude. As the aircraft climbs and the air density drops, the pilot must continuously open the throttle in small increments to maintain full power. The altitude at which the throttle reaches full open and the engine is still producing full rated power is known as the critical altitude. Above the critical altitude, engine power output will start to drop as the aircraft continues to climb.

Effects of Temperature

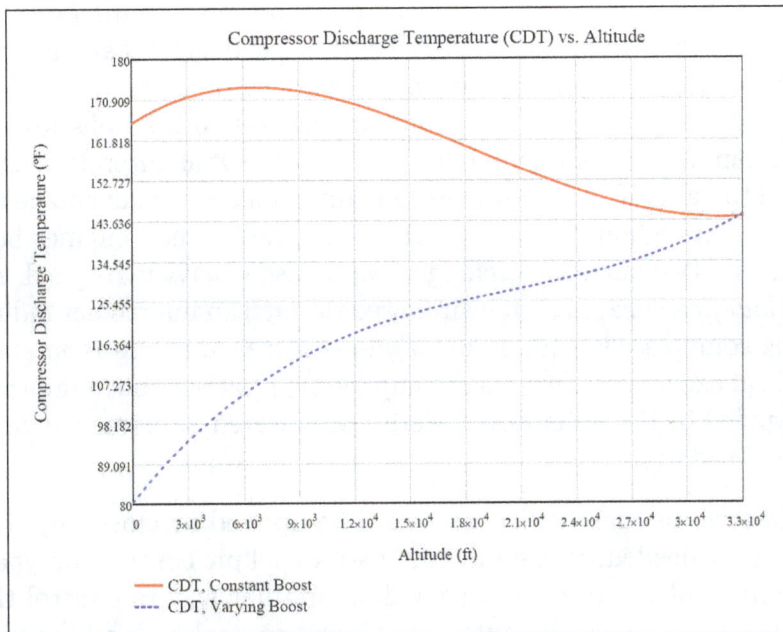

Supercharger CDT vs. altitude: Graph shows the CDT differences between a constant-boost supercharger and a variable-boost supercharger when utilized on an aircraft.

As discussed above, supercharging can cause a spike in temperature, and extreme temperatures will cause detonation of the fuel-air mixture and damage to the engine. In the case of aircraft, this causes a problem at low altitudes, where the air is both denser and warmer than at high altitudes. With high ambient air temperatures, detonation could start to occur with the manifold pressure gauge reading far below the red line.

A supercharger optimized for high altitudes causes the opposite problem on the intake side of the system. With the throttle retarded to avoid over-boosting, air temperature in the carburetor can drop low enough to cause ice to form at the throttle plate. In this manner, enough ice could accumulate to cause engine failure, even with the engine operating at full rated power. For this reason, many supercharged aircraft featured a carburetor air temperature gauge or warning light to alert the pilot of possible icing conditions.

Several solutions to these problems were developed: intercoolers and aftercoolers, anti-detonant injection, two-speed superchargers, and two-stage superchargers.

Two-speed and Two-stage Superchargers

In the 1930s, two-speed drives were developed for superchargers for aero engines providing more flexibility aircraft operation. The arrangement also entailed more complexity of manufacturing and maintenance. The gears connected the supercharger to the engine using a system of hydraulic clutches, which were initially manually engaged or disengaged by the pilot with a control in the cockpit. At low altitudes, the low-speed gear would be used in order to keep the manifold temperatures low. At around 12,000 feet (3,700 m), when the throttle was full forward and the manifold pressure started to drop off, the pilot would retard the throttle and switch to the higher gear, then readjust the throttle to the desired manifold pressure. Later installations automated the gear change according to atmospheric pressure.

In the Battle of Britain the Spitfire and Hurricane planes powered by the Rolls-Royce Merlin engine were equipped largely with single stage and single speed superchargers. Stanley Hooker of Rolls Royce, to improve the performance of the Merlin engine developed two-speed two-stage supercharging with aftercooling with a successful application on the Rolls Royce Merlin 61 aero engine in 1942. Horsepower was increased and performance at all aircraft heights. Hooker's developments allowed the aircraft they powered to maintain a crucial advantage over the German aircraft they opposed throughout World War II despite the German engines being significantly larger in displacement. Two-stage superchargers were also always two-speed. After the air was compressed in the *low-pressure stage*, the air flowed through an intercooler radiator where it was cooled before being compressed again by the *high-pressure stage* and then possibly also *aftercooled* in another heat exchanger. Two-stage compressors provided much improved high altitude performance, as typified by the Rolls-Royce Merlin 61 powered Supermarine Spitfire Mk IX and the North American Mustang.

In some two-stage systems, damper doors would be opened or closed by the pilot in order to bypass one stage as needed. Some systems had a cockpit control for opening or closing a damper to the intercooler/aftercooler, providing another way to control the temperature. Rolls-Royce Merlin engines had fully automated boost control with all the pilot having to do was advance the throttle with the control system limiting boost as necessary until maximum altitude was reached.

Turbocharging

A mechanically driven supercharger has to take its drive power from the engine. Taking a single-stage single-speed supercharged engine, such as an early Rolls-Royce Merlin, for instance, the supercharger uses up about 150 hp (110 kW). Without a supercharger, the engine could produce about 750 horsepower (560 kilowatts), but with a supercharger, it produces about 1,000 hp (750 kW)—an increase of about 400 hp (750 - 150 + 400 = 1000 hp), or a net gain of 250 hp (190 kW). This is where the principal disadvantage of a supercharger becomes apparent. The engine has to burn extra fuel to provide power to drive the supercharger. The increased air density during the input cycle increases the specific power of the engine and its power-to-weight ratio, but at the cost of an increase in the specific fuel consumption of the engine. In addition to increasing the cost of running the aircraft a supercharger has the potential to reduce its overall range for a specific fuel load.

As opposed to a supercharger driven by the engine itself, a turbocharger is driven using the otherwise wasted exhaust gas from the engine. The amount of power in the gas is proportional to the difference between the exhaust pressure and air pressure, and this difference increases with altitude, helping a turbocharged engine to compensate for changing altitude. This increases the height at which maximum power output of the engine is attained compared to supercharger boosting, and allows better fuel consumption at high altitude compared to an equivalent supercharged engine. This facilitates increased true airspeed at high altitude and gives a greater operational range than an equivalently boosted engine using a supercharger.

The majority of aircraft engines used during World War II used mechanically driven superchargers because they had some significant manufacturing advantages over turbochargers. However, the benefit to the operational range was given a much higher priority to American aircraft because of a less predictable requirement on the operational range and having to travel far from their home bases. Consequently, turbochargers were mainly employed in American aircraft engines such as the Allison V-1710 and the Pratt & Whitney R-2800, which were comparably heavier when turbocharged, and required additional ducting of expensive high-temperature metal alloys in the gas turbine and pre-turbine section of the exhaust system. The size of the ducting alone was a serious design consideration. For example, both the F4U Corsair and the P-47 Thunderbolt used the same radial engine, but the large barrel-shaped fuselage of the turbocharged P-47 was needed because of the amount of ducting to and from the turbocharger in the rear of the aircraft. The F4U used a two-stage intercooled supercharger with a more compact layout. Nonetheless, turbochargers were useful in high-altitude bombers and some fighter aircraft due to the increased high altitude performance and range.

Turbocharged piston engines are also subject to many of the same operating restrictions as those of gas turbine engines. Turbocharged engines also require frequent inspections of their turbochargers and exhaust systems to search for possible damage caused by the extreme heat and pressure of the turbochargers. Such damage was a prominent problem in the early models of the American Boeing B-29 Superfortress high-altitude bombers used in the Pacific Theater of Operations during 1944–45.

Turbocharged piston engines continued to be used in a large number of postwar airplanes, such as the B-50 Superfortress, the KC-97 Stratofreighter, the Boeing Stratoliner, the Lockheed Constellation, and the C-124 Globemaster II.

In more recent times most aircraft engines for general aviation (light airplanes) are naturally aspirated, but the smaller number of modern aviation piston engines designed to run at high altitudes use turbocharger or turbo-normalizer systems, instead of a supercharger driven from the crankshafts. The change in thinking is largely due to economics. Aviation gasoline was once plentiful and cheap, favoring the simple but fuel-hungry supercharger. As the cost of fuel has increased, the ordinary supercharger has fallen out of favor. Also, depending on what monetary inflation factor one uses, fuel costs have not decreased as fast as production and maintenance costs have.

Effects of Fuel Octane Rating

Until the late 1920s, all automobile and aviation fuel was generally rated at 87 octane or less. This is the rating that was achieved by the simple distillation of "light crude" oil. Engines from around the world were designed to work with this grade of fuel, which set a limit to the amount of boosting that could be provided by the supercharger while maintaining a reasonable compression ratio.

Octane rating boosting through additives was a line of research being explored at the time. Using these techniques, less valuable crude could still supply large amounts of useful gasoline, which made it a valuable economic process. However, the additives were not limited to making poor-quality oil into 87-octane gasoline; the same additives could also be used to boost the gasoline to much higher octane ratings.

Higher-octane fuel resists auto ignition and detonation better than does low-octane fuel. As a result, the amount of boost supplied by the superchargers could be increased, resulting in an increase in engine output. The development of 100-octane aviation fuel, pioneered in the USA before the war, enabled the use of higher boost pressures to be used on high-performance aviation engines and was used to develop extremely high-power outputs: for short periods: in several of the pre-war speed record airplanes. Operational use of the new fuel during World War II began in early 1940 when 100-octane fuel was delivered to the British Royal Air Force from refineries in America and the East Indies. The German *Luftwaffe* also had supplies of a similar fuel.

Increasing the knocking limits of existing aviation fuels became a major focus of aero engine development during World War II. By the end of the war, fuel was being delivered at a nominal 150-octane rating, on which late-war aero engines like the Rolls-Royce Merlin 66 or the Daimler-Benz DB 605DC developed as much as 2,000 hp (1,500 kW).

Turbocharger

A turbocharger, colloquially known as a turbo, is a turbine-driven forced induction device that increases an internal combustion engine's efficiency and power output by forcing extra compressed air into the combustion chamber. This improvement over a naturally aspirated engine's power output is due to the fact that the compressor can force more air—and proportionately more fuel—into the combustion chamber than atmospheric pressure (and for that matter, ram air intakes) alone.

Turbochargers were originally known as turbosuperchargers when all forced induction devices were classified as superchargers. Today the term "supercharger" is typically applied only to mechanically driven forced induction devices. The key difference between a turbocharger and a conventional supercharger is that a supercharger is mechanically driven by the engine, often through a belt connected to the crankshaft, whereas a turbocharger is powered by a turbine driven by the engine's exhaust gas. Compared with a mechanically driven supercharger, turbochargers tend to be more efficient, but less responsive. Twincharger refers to an engine with both a supercharger and a turbocharger.

Manufacturers commonly use turbochargers in truck, car, train, aircraft, and construction-equipment engines. They are most often used with Otto cycle and Diesel cycle internal combustion engines.

Cut-away view of an air foil bearing-supported turbocharger.

Turbocharging versus Supercharging

In contrast to turbochargers, superchargers are mechanically driven by the engine. Belts, chains, shafts, and gears are common methods of powering a supercharger, placing a mechanical load on the engine. For example, on the single-stage single-speed supercharged Rolls-Royce Merlin engine, the supercharger uses about 150 hp (110 kW). Yet the benefits outweigh the costs; for the 150 hp (110 kW) to drive the supercharger the engine generates an additional 400 hp (300 kW), a net gain of 250 hp (190 kW). This is where the principal disadvantage of a supercharger becomes apparent; the engine must withstand the net power output of the engine plus the power to drive the supercharger.

Another disadvantage of some superchargers is lower adiabatic efficiency when compared with turbochargers (especially Roots-type superchargers). Adiabatic efficiency is a measure of a compressor's ability to compress air without adding excess heat to that air. Even under ideal conditions, the compression process always results in elevated output temperature; however, more efficient compressors produce less excess heat. Roots superchargers impart significantly more heat to the air than turbochargers. Thus, for a given volume and pressure of air, the turbocharged air is cooler, and as a result denser, containing more oxygen molecules, and therefore more potential power than the supercharged air. In practical application the disparity between the two can be dramatic, with turbochargers often producing 15% to 30% more power based solely on the differences in adiabatic efficiency (however, due to heat transfer from the hot exhaust, considerable heating does occur).

By comparison, a turbocharger does not place a direct mechanical load on the engine, although turbochargers place exhaust back pressure on engines, increasing pumping losses. This is more efficient because while the increased back pressure taxes the piston exhaust stroke, much of the energy driving the turbine is provided by the still-expanding exhaust gas that would otherwise be wasted as heat through the tailpipe. In contrast to supercharging, the primary disadvantage of turbocharging is what is referred to as "lag" or "spool time". This is the time between the demand for an increase in power (the throttle being opened) and the turbochargers providing increased intake pressure, and hence increased power.

Throttle lag occurs because turbochargers rely on the buildup of exhaust gas pressure to drive the turbine. In variable output systems such as automobile engines, exhaust gas pressure at idle, low engine speeds, or low throttle is usually insufficient to drive the turbine. Only when the engine reaches sufficient speed does the turbine section start to *spool up,* or spin fast enough to produce intake pressure above atmospheric pressure.

A combination of an exhaust-driven turbocharger and an engine-driven supercharger can mitigate the weaknesses of both. This technique is called twincharging.

In the case of Electro-Motive Diesel's two-stroke engines, the mechanically assisted turbocharger is not specifically a twincharger, as the engine uses the mechanical assistance to charge air only at lower engine speeds and startup. Once above notch 5, the engine uses true turbocharging. This differs from a turbocharger that uses the compressor section of the turbo-compressor only during starting and, as a two-stroke engines cannot naturally aspirate, and, according to SAE definitions, a two-stroke engine with a mechanically assisted compressor during idle and low throttle is considered naturally aspirated.

Operating Principle

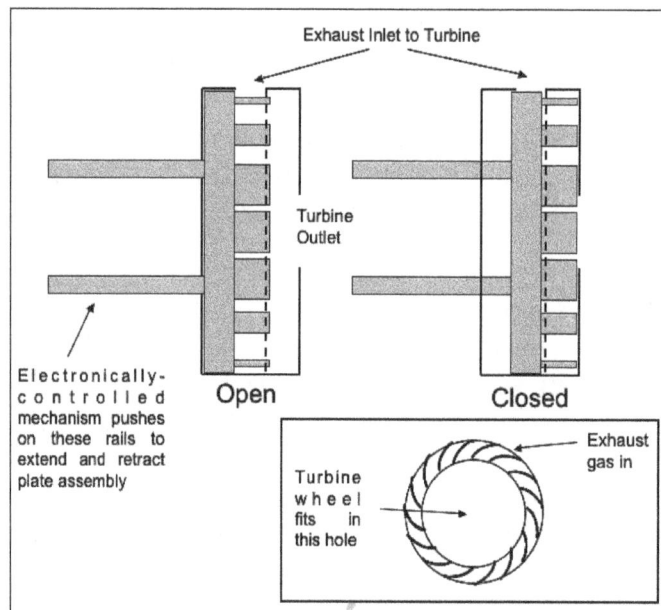

In naturally aspirated piston engines, intake gases are drawn or "pushed" into the engine by atmospheric pressure filling the volumetric void caused by the downward stroke of the piston (which creates a low-pressure area), similar to drawing liquid using a syringe. The amount of air actually inspired, compared with the theoretical amount if the engine could maintain atmospheric pressure, is called volumetric efficiency. The objective of a turbocharger is to improve an engine's volumetric efficiency by increasing density of the intake gas (usually air) allowing more power per engine cycle.

The turbocharger's compressor draws in ambient air and compresses it before it enters into the intake manifold at increased pressure. This results in a greater mass of air entering the cylinders on each intake stroke. The power needed to spin the centrifugal compressor is derived from the kinetic energy of the engine's exhaust gases.

In automotive applications, 'boost' refers to the amount by which intake manifold pressure exceeds atmospheric pressure. This is representative of the extra air pressure that is achieved over what would be achieved without the forced induction. The level of boost may be shown on a pressure gauge, usually in bar, psi or possibly kPa. The control of turbocharger boost has changed dramatically over the 100-plus years of their use.

In petrol engine turbocharger applications, boost pressure is limited to keep the entire engine system, including the turbocharger, inside its thermal and mechanical design operating range. Over-boosting an engine frequently causes damage to the engine in a variety of ways including pre-ignition, overheating, and over-stressing the engine's internal hardware. For example, to avoid engine knocking (also known as detonation) and the related physical damage to the engine, the intake manifold pressure must not get too high, thus the pressure at the intake manifold of the engine must be controlled by some means. Opening the wastegate allows the excess energy destined for the turbine to bypass it and pass directly to the exhaust pipe, thus reducing boost pressure. The wastegate can be either controlled manually or by an actuator (in automotive applications, it is often controlled by the engine control unit).

Pressure Increase (or Boost)

A turbocharger may also be used to increase fuel efficiency without increasing power. This is achieved by diverting exhaust waste energy, from the combustion process, and feeding it back into the turbo's "hot" intake side that spins the turbine. As the hot turbine side is being driven by the exhaust energy, the cold intake turbine (the other side of the turbo) compresses fresh intake air and drives it into the engine's intake. By using this otherwise wasted energy to increase the mass of air, it becomes easier to ensure that all fuel is burned before being vented at the start of the exhaust stage. The increased temperature from the higher pressure gives a higher Carnot efficiency.

A reduced density of intake air is caused by the loss of atmospheric density seen with elevated altitudes. Thus, a natural use of the turbocharger is with aircraft engines. As an aircraft climbs to higher altitudes, the pressure of the surrounding air quickly falls off. At 18,000 feet (5,500 m), the air is at half the pressure of sea level, which means that the engine produces less than half-power at this altitude. In aircraft engines, turbocharging is commonly used to maintain manifold pressure as altitude increases (i.e. to compensate for lower-density air at higher altitudes). Since atmospheric pressure reduces as the aircraft climbs, power drops as a function of altitude in normally aspirated engines. Systems that use a turbocharger to maintain an engine's sea-level power output are called turbo-normalized systems. Generally, a turbo-normalized system attempts to maintain a manifold pressure of 29.5 inHg (100 kPa).

Turbocharger Lag

Turbocharger lag (turbo lag) is the time required to change power output in response to a throttle change, noticed as a hesitation or slowed *throttle response* when accelerating as compared to a naturally aspirated engine. This is due to the time needed for the exhaust system and turbocharger to generate the required boost which can also be referred to as spooling. Inertia, friction, and compressor load are the primary contributors to turbocharger lag. Superchargers do not suffer this problem, because the turbine is eliminated due to the compressor being directly powered by the engine.

Turbocharger applications can be categorized into those that require changes in output power (such as automotive) and those that do not (such as marine, aircraft, commercial automotive, industrial, engine-generators, and locomotives). While important to varying degrees, turbocharger lag is most problematic in applications that require rapid changes in power output. Engine designs reduce lag in a number of ways:

- Lowering the rotational inertia of the turbocharger by using lower radius parts and ceramic and other lighter materials.

- Changing the turbine's *aspect ratio*.

- Increasing upper-deck air pressure (compressor discharge) and improving wastegate response.

- Reducing bearing frictional losses, e.g., using a foil bearing rather than a conventional oil bearing.

- Using variable-nozzle or twin-scroll turbochargers.

- Decreasing the volume of the upper-deck piping.

- Using multiple turbochargers sequentially or in parallel.

- Using an antilag system.

- Using a turbocharger spool valve to increase exhaust gas flow speed to the (twin-scroll) turbine.

Sometimes turbo lag is mistaken for engine speeds that are below boost threshold. If engine speed is below a turbocharger's boost threshold rpm then the time needed for the vehicle to build speed and rpm could be considerable, maybe even tens of seconds for a heavy vehicle starting at low vehicle speed in a high gear. This wait for vehicle speed increase is not turbo lag, it is improper gear selection for boost demand. Once the vehicle reaches sufficient speed to provide the required rpm to reach boost threshold, there will be a far shorter delay while the turbo itself builds rotational energy and transitions to positive boost, only this last part of the delay in achieving positive boost is the turbo lag.

Boost Threshold

The *boost threshold* of a turbocharger system is the lower bound of the region within which the compressor operates. Below a certain rate of flow, a compressor produces insignificant boost. This limits boost at a particular RPM, regardless of exhaust gas pressure. Newer turbocharger and engine developments have steadily reduced boost thresholds.

Electrical boosting ("E-boosting") is a new technology under development. It uses an electric motor to bring the turbocharger up to operating speed quicker than possible using available exhaust gases. An alternative to e-boosting is to completely separate the turbine and compressor into a turbine-generator and electric-compressor as in the hybrid turbocharger. This makes compressor speed independent of turbine speed. In 1981, a similar system that used a hydraulic drive system and overspeed clutch arrangement accelerated the turbocharger of the MV *Canadian Pioneer* (Doxford 76J4CR engine).

Turbochargers start producing boost only when a certain amount of kinetic energy is present in the exhaust gasses. Without adequate exhaust gas flow to spin the turbine blades, the turbocharger cannot produce the necessary force needed to compress the air going into the engine. The boost threshold is determined by the engine displacement, engine rpm, throttle opening, and the size of the turbocharger. The operating speed (rpm) at which there is enough exhaust gas momentum to compress the air going into the engine is called the "boost threshold rpm". Reducing the "boost threshold rpm" can improve throttle response.

Key Components

The turbocharger has three main components:

- The turbine, which is almost always a radial inflow turbine (but is almost always a single-stage axial inflow turbine in large Diesel engines).

- The compressor, which is almost always a centrifugal compressor.

- The center housing/hub rotating assembly.

Many turbocharger installations use additional technologies, such as wastegates, intercooling and blow-off valves.

Turbine

Energy provided for the turbine work is converted from the enthalpy and kinetic energy of the gas. The turbine housings direct the gas flow through the turbine as it spins at up to 250,000 rpm. The size and shape can dictate some performance characteristics of the overall turbocharger. Often the same basic turbocharger assembly is available from the manufacturer with multiple housing choices for the turbine, and sometimes the compressor cover as well. This lets the balance between performance, response, and efficiency be tailored to the application.

On the left, the brass oil drain connection. On the right are the braided oil supply line and water coolant line connections.

Compressor impeller side with the cover removed.

Turbine side housing removed.

The turbine and impeller wheel sizes also dictate the amount of air or exhaust that can flow through the system, and the relative efficiency at which they operate. In general, the larger the turbine wheel and compressor wheel the larger the flow capacity. Measurements and shapes can vary, as well as curvature and number of blades on the wheels.

A turbocharger's performance is closely tied to its size. Large turbochargers take more heat and pressure to spin the turbine, creating lag at low speed. Small turbochargers spin quickly, but may not have the same performance at high acceleration. To efficiently combine the benefits of large and small wheels, advanced schemes are used such as twin-turbochargers, twin-scroll turbochargers, or variable-geometry turbochargers.

Twin-turbo

Twin-turbo or bi-turbo designs have two separate turbochargers operating in either a sequence or in parallel. In a parallel configuration, both turbochargers are fed one-half of the engine's exhaust. In a sequential setup one turbocharger runs at low speeds and the second turns on at a predetermined engine speed or load. Sequential turbochargers further reduce turbo lag, but require an intricate set of pipes to properly feed both turbochargers.

Two-stage variable twin-turbos employ a small turbocharger at low speeds and a large one at higher speeds. They are connected in a series so that boost pressure from one turbocharger is multiplied by another, hence the name "2-stage." The distribution of exhaust gas is continuously variable, so the transition from using the small turbocharger to the large one can be done incrementally. Twin turbochargers are primarily used in Diesel engines. For example, in Opel bi-turbo Diesel, only the smaller turbocharger works at low speed, providing high torque at 1,500–1,700 rpm. Both turbochargers operate together in mid range, with the larger one pre-compressing the air, which the smaller one further compresses. A bypass valve regulates the exhaust flow to each turbocharger. At higher speed (2,500 to 3,000 RPM) only the larger turbocharger runs.

Smaller turbochargers have less turbo lag than larger ones, so often two small turbochargers are used instead of one large one. This configuration is popular in engines over 2.5-litres and in V-shape or boxer engines.

Twin-scroll

Twin-scroll or divided turbochargers have two exhaust gas inlets and two nozzles, a smaller sharper angled one for quick response and a larger less angled one for peak performance.

With high-performance camshaft timing, exhaust valves in different cylinders can be open at the same time, overlapping at the end of the power stroke in one cylinder and the end of exhaust stroke in another. In twin-scroll designs, the exhaust manifold physically separates the channels for cylinders that can interfere with each other, so that the pulsating exhaust gasses flow through separate spirals (scrolls). With common firing order 1–3–4–2, two scrolls of unequal length pair cylinders 1 and 4, and 3 and 2. This lets the engine efficiently use exhaust scavenging techniques, which decreases exhaust gas temperatures and NO_x emissions, improves turbine efficiency, and reduces turbo lag evident at low engine speeds.

Cut-out of a twin-scroll turbocharger, with two differently angled nozzles.

Cut-out of a twin-scroll exhaust and turbine; the dual "scrolls" pairing cylinders 1 and 4, and 2 and 3 are clearly visible.

Variable-geometry

Variable-geometry or variable-nozzle turbochargers use moveable vanes to adjust the air-flow to the turbine, imitating a turbocharger of the optimal size throughout the power curve. The vanes are placed just in front of the turbine like a set of slightly overlapping walls. Their angle is adjusted by an actuator to block or increase air flow to the turbine. This variability maintains a comparable exhaust velocity and back pressure throughout the engine's rev range. The result is that the turbocharger improves fuel efficiency without a noticeable level of turbocharger lag.

Garrett variable-geometry turbocharger on DV6TED4 engine.

Compressor

The compressor increases the mass of intake air entering the combustion chamber. The compressor is made up of an impeller, a diffuser and a volute housing.

The operating range of a compressor is described by the "compressor map".

Ported Shroud

The flow range of a turbocharger compressor can be increased by allowing air to bleed from a ring of holes or a circular groove around the compressor at a point slightly downstream of the compressor inlet (but far nearer to the inlet than to the outlet).

The ported shroud is a performance enhancement that allows the compressor to operate at significantly lower flows. It achieves this by forcing a simulation of impeller stall to occur continuously. Allowing some air to escape at this location inhibits the onset of surge and widens the operating range. While peak efficiencies may decrease, high efficiency may be achieved over a greater range of engine speeds. Increases in compressor efficiency result in slightly cooler (more dense) intake air, which improves power. This is a passive structure that is constantly open (in contrast to compressor exhaust blow off valves, which are mechanically or electronically controlled). The ability of the compressor to provide high boost at low rpm may also be increased marginally (because near choke conditions the compressor draws air inward through the bleed path). Ported shrouds are used by many turbocharger manufacturers.

Center Housing/Hub Rotating Assembly

The centre hub rotating assembly (CHRA) houses the shaft that connects the compressor impeller and turbine. It also must contain a bearing system to suspend the shaft, allowing it to rotate at very high speed with minimal friction. For instance, in automotive applications the CHRA typically uses a thrust bearing or ball bearing lubricated by a constant supply of pressurized engine oil. The CHRA may also be considered "water-cooled" by having an entry and exit point for engine coolant. Water-cooled models use engine coolant to keep lubricating oil cooler, avoiding possible oil coking (destructive distillation of engine oil) from the extreme heat in the turbine. The development of air-foil bearings removed this risk.

Ball bearings designed to support high speeds and temperatures are sometimes used instead of fluid bearings to support the turbine shaft. This helps the turbocharger accelerate more quickly and reduces turbo lag. Some variable nozzle turbochargers use a rotary electric actuator, which uses a direct stepper motor to open and close the vanes, rather than pneumatic controllers that operate based on air pressure.

Additional Technologies Commonly used in Turbocharger Installations

Intercooling

When the pressure of the engine's intake air is increased, its temperature also increases. This occurrence can be explained through Gay-Lussac's law, stating that the pressure of a given amount of gas held at constant volume is directly proportional to the Kelvin temperature. With more pressure being added to the engine through the turbocharger, overall temperatures of the engine will also rise. In addition, heat soak from the hot exhaust gases spinning the turbine will also heat the intake air. The warmer the intake air, the less dense, and the less oxygen available for the combustion event, which reduces volumetric efficiency. Not only does excessive intake-air temperature reduce efficiency, it also leads to engine knock, or detonation, which is destructive to engines.

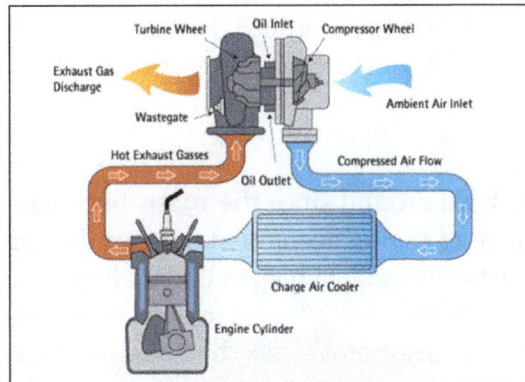

Illustration of typical component layout in a production turbocharged petrol engine.

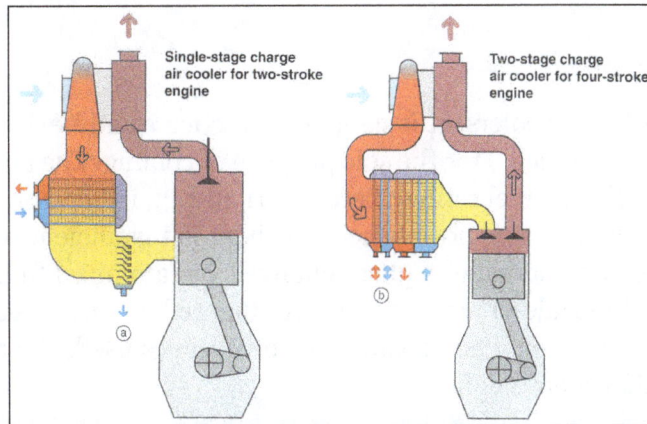

Illustration of inter-cooler location.

To compensate for the increase in temperature, turbocharger units often make use of an intercooler between successive stages of boost to cool down the intake air. A *charge air cooler* is an air cooler between the boost stages and the appliance that consumes the boosted air.

Top-mount (TMIC) vs. Front-mount Intercoolers (FMIC)

There are two areas on which intercoolers are commonly mounted. It can be either mounted on top, parallel to the engine, or mounted near the lower front of the vehicle. Top-mount intercoolers setups will result in a decrease in turbo lag, due in part by the location of the intercooler being much closer to the turbocharger outlet and throttle body. This closer proximity reduces the time it takes for air to travel through the system, producing power sooner, compared to that of a front-mount intercooler which has more distance for the air to travel to reach the outlet and throttle.

Front-mount intercoolers can have the potential to give better cooling compared to that of a top-mount. The area in which a top-mounted intercooler is located, is near one of the hottest areas of a car, right above the engine. This is why most manufacturers include large hood scoops to help feed air to the intercooler while the car is moving, but while idle, the hood scoop provides little to no benefit. Even while moving, when the atmospheric temperatures begin to rise, top-mount intercoolers tend to underperform compared to front-mount intercoolers. With more distance to travel, the air circulated through a front-mount intercooler may have more time to cool.

Water Injection

An alternative to intercooling is injecting water into the intake air to reduce the temperature. This method has been used in automotive and aircraft applications.

Methanol Injection

Methanol/water injection has been around since the 1920s but was not utilized until World War II. Adding the mixture to intake of the turbocharged engines decreased operating temperatures and increased horse power. Turbocharged engines today run high boost and high engine temperatures to match. When injecting the mixture into the intake stream, the air is cooled as the liquids evaporate. Inside the combustion chamber it slows the flame, acting similar to higher octane fuel. Methanol/water mixture allows for higher compression because of the less detonation-prone and, thus, safer combustion inside the engine.

Fuel-air Mixture Ratio

In addition to the use of intercoolers, it is common practice to add extra fuel to the intake air (known as "running an engine rich") for the sole purpose of cooling. The amount of extra fuel varies, but typically reduces the air-fuel ratio to between 11 and 13, instead of the stoichiometric 14.7 (in petrol engines). The extra fuel is not burned (as there is insufficient oxygen to complete the chemical reaction), instead it undergoes a phase change from atomized (liquid) to gas. This phase change absorbs heat, and the added mass of the extra fuel reduces the average thermal energy of the charge and exhaust gas. Even when a catalytic converter is used, the practice of running an engine rich increases exhaust emissions.

Wastegate

A wastegate regulates the exhaust gas flow that enters the exhaust-side driving turbine and therefore the air intake into the manifold and the degree of boosting. It can be controlled by a boost pressure assisted, generally vacuum hose attachment point diaphragm (for vacuum and posi-tive pressure to return commonly oil contaminated waste to the emissions system) to force the spring-loaded diaphragm to stay closed until the overboost point is sensed by the ecu or a solenoid operated by the engine's electronic control unit or a boost controller, but most production vehicles use a single vacuum hose attachment point spring-loaded diaphragm that can alone be pushed open, thus limiting overboost ability due to exhaust gas pressure forcing open the wastegate.

Anti-surge/Dump/blow off Valves

A recirculating type anti-surge valve.

Turbocharged engines operating at wide open throttle and high rpm require a large volume of air to flow between the turbocharger and the inlet of the engine. When the throttle is closed, com-pressed air flows to the throttle valve without an exit (i.e., the air has nowhere to go).

In this situation, the surge can raise the pressure of the air to a level that can cause damage. This is because if the pressure rises high enough, a compressor stall occurs—stored pressurized air de-compresses backward across the impeller and out the inlet. The reverse flow back across the tur-bocharger makes the turbine shaft reduce in speed more quickly than it would naturally, possibly damaging the turbocharger.

To prevent this from happening, a valve is fitted between the turbocharger and inlet, which vents off the excess air pressure. These are known as an anti-surge, diverter, bypass, turbo-relief valve, blow-off valve (BOV), or dump valve. It is a pressure relief valve, and is normally operated by the vacuum from the intake manifold.

The primary use of this valve is to maintain the spinning of the turbocharger at a high speed. The air is usually recycled back into the turbocharger inlet (diverter or bypass valves), but can also be vented to the atmosphere (blow off valve). Recycling back into the turbocharger inlet is required on an engine that uses a mass-airflow fuel injection system, because dumping the excessive air overboard downstream of the mass airflow sensor causes an excessively rich fuel mixture—be-cause the mass-airflow sensor has already accounted for the extra air that is no longer being used. Valves that recycle the air also shorten the time needed to re-spool the turbocharger after sudden engine deceleration, since load on the turbocharger when the valve is active is much lower than if the air charge vents to atmosphere.

Free Floating

A free floating turbocharger is used in the 100-litre engine of this Caterpillar mining vehicle.

A free floating turbocharger is the simplest type of turbocharger. This configuration has no wastegate and cannot control its own boost levels. They are typically designed to attain maximum boost at full throttle. Free floating turbochargers produce more horsepower because they have less backpressure, but are not driveable in performance applications without an external wastegate.

Applications

Petrol-powered Cars

The first turbocharged passenger car was the Oldsmobile Jetfire option on the 1962–1963 F85/ Cutlass, which used a turbocharger mounted to a 215 cu in (3.52 L) all aluminum V8. Also in 1962, Chevrolet introduced a special run of turbocharged Corvairs, initially called the Monza Spyder (1962–1964) and later renamed the Corsa (1965–1966), which mounted a turbocharger to its air cooled flat six cylinder engine. This model popularized the turbocharger in North America—and set the stage for later turbocharged models from Porsche on the 1975-up 911/930, Saab on the 1978–1984 Saab 99 Turbo, and the very popular 1978–1987 Buick Regal/T Type/Grand National. Today, turbocharging is common on both diesel and petrol-powered cars. Turbocharging can increase power output for a given capacity or increase fuel efficiency by allowing a smaller displacement engine. The 'Engine of the year 2011' is an engine used in a Fiat 500 equipped with an MHI turbocharger. This engine lost 10% weight, saving up to 30% in fuel consumption while delivering the same HP (105) as a 1.4-litre engine.

Diesel-powered Cars

The first production turbocharger diesel passenger car was the Garrett-turbocharged Mercedes 300SD introduced in 1978. Today, most automotive diesels are turbocharged, since the use of turbocharging improved efficiency, driveability and performance of diesel engines, greatly increasing their popularity. The Audi R10 with a diesel engine even won the 24 hours race of Le Mans in 2006, 2007 and 2008.

Motorcycles

The first example of a turbocharged bike is the 1978 Kawasaki Z1R TC. Several Japanese companies produced turbocharged high-performance motorcycles in the early 1980s, such as the CX500 Turbo from Honda- a transversely mounted, liquid cooled V-Twin also available in naturally aspirated form. Since then, few turbocharged motorcycles have been produced. This is partially due to an abundance of larger displacement, naturally aspirated engines being available that offer the torque and power benefits of a smaller displacement engine with turbocharger, but do return more linear power characteristics. The Dutch manufacturer EVA motorcycles builds a small series of turbocharged diesel motorcycle with an 800cc smart CDI engine.

Trucks

The first turbocharged diesel truck was produced by *Schweizer Maschinenfabrik Saurer* (Swiss Machine Works Saurer) in 1938.

Aircraft

A natural use of the turbocharger—and its earliest known use for any internal combustion engine, starting with experimental installations in the 1920s—is with aircraft engines. As an aircraft climbs to higher altitudes the pressure of the surrounding air quickly falls off. At 5,486 m (18,000 ft), the air is at half the pressure of sea level and the airframe experiences only half the aerodynamic drag. However, since the charge in the cylinders is pushed in by this air pressure, the engine normally produces only half-power at full throttle at this altitude. Pilots would like to take advantage of the low drag at high altitudes to go faster, but a naturally aspirated engine does not produce enough power at the same altitude to do so.

The table below is used to demonstrate the wide range of conditions experienced. As seen in the table below, there is significant scope for forced induction to compensate for lower density environments.

	Daytona Beach	Denver	Death Valley	Colorado State Highway 5	La Rinconada, Peru,
elevation	0 m / 0 ft	1,609 m / 280 ft	−86 m / −282 ft	4,347 m / 14,264 ft	5,100 m / 16,732 ft
atm	1.000	0.823	1.010	0.581	0.526
bar	1.013	0.834	1.024	0.589	0.533
psia	14.696	12.100	14.846	8.543	7.731
kPa	101.3	83.40	102.4	58.90	53.30

A turbocharger remedies this problem by compressing the air back to sea-level pressures (turbo-normalizing), or even much higher (turbo-charging), in order to produce rated power at high altitude. Since the size of the turbocharger is chosen to produce a given amount of pressure at high altitude, the turbocharger is oversized for low altitude. The speed of the turbocharger is controlled by a wastegate. Early systems used a fixed wastegate, resulting in a turbocharger that functioned much like a supercharger. Later systems utilized an adjustable wastegate, controlled either manually by the pilot or by an automatic hydraulic or electric system. When the aircraft is at low altitude the wastegate is usually fully open, venting all the exhaust gases overboard. As the aircraft climbs and the air density drops, the wastegate must continuously close in small increments to maintain full power. The altitude at which the wastegate fully closes and the engine still produces full power is the *critical altitude*. When the aircraft climbs above the critical altitude, engine power output decreases as altitude increases, just as it would in a naturally aspirated engine.

With older supercharged aircraft without Automatic Boost Control, the pilot must continually adjust the throttle to maintain the required manifold pressure during ascent or descent. The pilot must also take care to avoid over-boosting the engine and causing damage. In contrast, modern turbocharger systems use an automatic wastegate, which controls the manifold pressure within parameters preset by the manufacturer. For these systems, as long as the control system is working properly and the pilot's control commands are smooth and deliberate, a turbocharger cannot overboost the engine and damage it.

Yet the majority of World War II engines used superchargers, because they maintained three significant manufacturing advantages over turbochargers, which were larger, involved extra piping, and required exotic high-temperature materials in the turbine and pre-turbine section of the exhaust system. The size of the piping alone is a serious issue; American fighters Vought F4U and Republic P-47 used the same engine, but the huge barrel-like fuselage of the latter was, in part, needed to hold the piping to and from the turbocharger in the rear of the plane. Turbocharged piston engines are also subject to many of the same operating restrictions as gas turbine engines. Pilots must make smooth, slow throttle adjustments to avoid overshooting their target manifold pressure. The fuel/air mixture must often be adjusted far on the rich side of stoichiometric combustion needs to avoid pre-ignition or detonation in the engine when running at high power settings. In systems using a manually operated wastegate, the pilot must be careful not to exceed the turbocharger's maximum rpm. The additional systems and piping increase an aircraft engine's size, weight, complexity and cost. A turbocharged aircraft engine costs more to maintain than a comparable normally aspirated engine. The great majority of World War II American heavy bombers used by the USAAF, particularly the Wright R-1820 *Cyclone-9* powered B-17 Flying Fortress, and Pratt & Whitney R-1830 Twin Wasp powered Consolidated B-24 Liberator four-engine bombers both used similar models of General Electric-designed turbochargers in service, as did the twin Allison V-1710-engined Lockheed P-38 Lightning American heavy fighter during the war years.

All of the above WWII aircraft engines had mechanically driven centrifugal superchargers as-designed from the start, and the turbosuperchargers (with intercoolers) were added, effectively as twincharger systems, to achieve desired altitude performance.

Today, most general aviation piston engine powered aircraft are naturally aspirated. Modern aviation piston engines designed to run at high altitudes typically include a turbocharger (either high pressure or turbonormalized) rather than a supercharger. The change in thinking is largely due to economics. Avgas was once plentiful and cheap, favouring the simple, but fuel-hungry, supercharger. As the cost of fuel has increased, the supercharger has fallen out of favour.

Turbocharged aircraft often occupy a performance range between that of normally aspirated piston-powered aircraft and turbine-powered aircraft. Despite the negative points, turbocharged aircraft fly higher for greater efficiency. High cruise flight also allows more time to evaluate issues before a forced landing must be made.

As the turbocharged aircraft climbs, however, the pilot (or automated system) can close the wastegate, forcing more exhaust gas through the turbocharger turbine, thereby maintaining manifold pressure during the climb, at least until the critical pressure altitude is reached (when the wastegate is fully closed), after which manifold pressure falls. With such systems, modern high-performance piston engine aircraft can cruise at altitudes up to 25,000 feet (above which, RVSM certification would be required), where low air density results in lower drag and higher true airspeeds. This allows flying "above the weather". In manually controlled wastegate systems, the pilot must take care not to overboost the engine, which causes detonation, leading to engine damage.

Marine and Land-based Diesel Turbochargers

A medium-sized six-cylinder marine diesel-engine, with turbocharger and exhaust in the foreground.

Turbocharging, which is common on diesel engines in automobiles, trucks, tractors, and boats is also common in heavy machinery such as locomotives, ships, and auxiliary power generation.

- Turbocharging can dramatically improve an engine's specific power and power-to-weight ratio, performance characteristics that are normally poor in non-turbocharged diesel engines.

- Diesel engines have no detonation because diesel fuel is injected at or towards the end of the compression stroke and is ignited solely by the heat of compression of the charge air. Because of this, diesel engines can use a much higher boost pressure than spark ignition engines, limited only by the engine's ability to withstand the additional heat and pressure.

Turbochargers are also employed in certain two-stroke cycle diesel engines, which would normally require a Roots blower for aspiration. In this specific application, mainly Electro-Motive Diesel (EMD) 567, 645, and 710 Series engines, the turbocharger is initially driven by the engine's crankshaft through a gear train and an overrunning clutch, thereby providing aspiration for combustion. After combustion has been achieved, and after the exhaust gases have reached sufficient heat energy, the overrunning clutch is automatically disengaged, and the turbo-compressor is thereafter driven exclusively by the exhaust gases. In the EMD application, the turbocharger acts as a compressor for normal aspiration during starting and low power output settings and is used for true turbocharging during medium and high power output settings. This is particularly beneficial at high altitudes, as are often encountered on western U.S. railroads. It is possible for the turbocharger to revert to compressor mode momentarily during commands for large increases in engine power.

Safety

Turbocharger failures and resultant high exhaust temperatures are among the causes of car fires.

Air Filter

The combustion air filter prevents abrasive particulate matter from entering the engine's cylinders, where it would cause mechanical wear and oil contamination.

Most fuel injected vehicles use a pleated paper filter element in the form of a flat panel. This filter is usually placed inside a plastic box connected to the throttle body with duct work. Older vehicles that use carburetors or throttle body fuel injection typically use a cylindrical air filter, usually between 100 millimetres (4 in) and 400 millimetres (16 in) in diameter. This is positioned above or beside the carburetor or throttle body, usually in a metal or plastic container which may incorporate ducting to provide cool and/or warm inlet air, and secured with a metal or plastic lid. The overall unit (filter and housing together) is called the air cleaner.

Used auto engine air filter, clean side.

Used auto engine air filter, dirty side.

Auto engine air filter clogged with dust and grime.

Low-temperature oxidation catalyst used to convert carbon monoxide to less toxic carbon dioxide at room temperature. It can also remove formaldehyde from the air.

Paper

Pleated paper filter elements are the nearly exclusive choice for automobile engine air cleaners, because they are efficient, easy to service, and cost-effective. The "paper" term is somewhat misleading, as the filter media are considerably different from papers used for writing or packaging, etc. There is a persistent belief among tuners, fomented by advertising for aftermarket non-paper replacement filters, that paper filters flow poorly and thus restrict engine performance. In fact, as long as a pleated-paper filter is sized appropriately for the airflow volumes encountered in a particular application, such filters present only trivial restriction to flow until the filter has become significantly clogged with dirt. Construction equipment engines also use this. The reason is that the paper is bent in zig-zag shape, and the total area of the paper is very large, in the range of 50 times of the air opening.

Foam

Oil-wetted polyurethane foam elements are used in some aftermarket replacement automobile air filters. Foam was in the past widely used in air cleaners on small engines on lawnmowers and other power equipment, but automotive-type paper filter elements have largely supplanted oil-wetted foam in these applications. Foam filters are still commonly used on air compressors for air tools up to 5Hp. Depending on the grade and thickness of foam employed, an oil-wetted foam filter element can offer minimal airflow restriction or very high dirt capacity, the latter property making foam filters a popular choice in off-road rallying and other motorsport applications where high levels of dust will be encountered. Due to the way dust is captured on foam filters, large amounts may be trapped without measurable change in airflow restriction.

Cotton

Oiled cotton gauze is employed in a growing number of aftermarket automotive air filters marketed as high-performance items. In the past, cotton gauze saw limited use in original-equipment automotive air filters. However, since the introduction of the Abarth SS versions, the Fiat subsidiary supplies cotton gauze air filters as OE filters.

Stainless Steel

Stainless steel mesh is another example of medium which allow more air to pass through. Stainless steel mesh comes with different mesh counts, offering different filtration standards. In an extreme modified engine lacking in space for a cone based air filter, some will opt to install a simple stainless steel mesh over the turbo to ensure no particles enter the engine via the turbo.

Oil Bath

An oil bath air cleaner consists of a sump containing a pool of oil, and an insert which is filled with fiber, mesh, foam, or another coarse filter media. When the cleaner is assembled, the media-containing body of the insert sits a short distance above the surface of the oil pool. The rim of the insert overlaps the rim of the sump. This arrangement forms a labyrinthine path through which the air must travel in a series of U-turns: up through the gap between the rims of the insert and the sump, down through the gap between the outer wall of the insert and the inner wall of the sump, and up through the filter media in the body of the insert. This U-turn takes the air at high velocity across the surface of the oil pool. Larger and heavier dust and dirt particles in the air cannot make the turn due to their inertia, so they fall into the oil and settle to the bottom of the base bowl. Lighter and smaller particles are trapped by the filtration media in the insert, which is wetted by oil droplets aspirated there into by normal airflow.

Oil bath air cleaners were very widely used in automotive and small engine applications until the widespread industry adoption of the paper filter in the early 1960s. Such cleaners are still used in off-road equipment where very high levels of dust are encountered, for oil bath air cleaners can sequester a great deal of dirt relative to their overall size without loss of filtration efficiency or airflow. However, the liquid oil makes cleaning and servicing such air cleaners messy and inconvenient, they must be relatively large to avoid excessive restriction at high airflow rates, and they tend to increase exhaust emissions of unburned hydrocarbons due to oil aspiration when used on spark-ignition engines.

Water Bath

In the early 20th century (about 1900 to 1930), water bath air cleaners were used in some applications (cars, trucks, tractors, and portable and stationary engines). They worked on roughly the same principles as oil bath air cleaners. For example, the original Fordson tractor had a water bath air cleaner. By the 1940s, oil bath designs had displaced water bath designs because of better filtering performance.

Manifold Absolute Pressure Sensor

The manifold absolute pressure sensor (MAP sensor) is one of the sensors used in an internal combustion engine's electronic control system.

Engines that use a MAP sensor are typically fuel injected. The manifold absolute pressure sensor provides instantaneous manifold pressure information to the engine's electronic control unit (ECU). The data is used to calculate air density and determine the engine's air mass flow rate, which in turn determines the required fuel metering for optimum combustion and influence the advance or retard of ignition timing. A fuel-injected engine may alternatively use a mass airflow sensor (MAF sensor) to detect the intake airflow. A typical naturally aspirated engine configuration employs one or the other, whereas forced induction engines typically use both; a MAF sensor on the charge pipe leading to the throttle body and a MAP sensor on the intake tract pre-turbo.

MAP sensor data can be converted to air mass data by using a second variable coming from an IAT Sensor (intake air temperature sensor). This is called the speed-density method. Engine speed (RPM) is also used to determine where on a look up table to determine fuelling, hence speed-density (engine speed / air density). The MAP sensor can also be used in OBD II (on-board diagnostics) applications to test the EGR (exhaust gas recirculation) valve for functionality, an application typical in OBD II equipped General Motors engines.

The following example assumes the same engine speed and air temperature.

- Condition 1:

An engine operating at wide open throttle (WOT) on top of a very high mountain has a manifold pressure of about 50 kPa (essentially equal to the barometer at that high altitude).

- Condition 2:

The same engine at sea level will achieve that same 50 kPa (7.25 psi, 14.7 inHG) of manifold pressure at less than (before reaching) WOT due to the higher barometric pressure.

The engine requires the same mass of fuel in both conditions because the mass of air entering the cylinders is the same.

If the throttle is opened all the way in condition 2, the manifold absolute pressure will increase from 50 kPa to nearly 100 kPa (14.5 psi, 29.53 inHG), about equal to the local barometer, which in condition 2 is sea level. The higher absolute pressure in the intake manifold increases the air's density, and in turn more fuel can be burned resulting in higher output.

Another example is varying rpm and engine loads:

Where an engine may have 60kPa of manifold pressure at 1800 rpm in an unloaded condition, introducing load with a further throttle opening will change the final manifold pressure to 100kPa, engine will still be at 1800 rpm but its loading will require a different spark and fueling delivery.

Vacuum Comparison

Engine vacuum is the difference between the pressures in the intake manifold and ambient atmospheric pressure. Engine vacuum is a "gauge" pressure, since gauges by nature measure a pressure difference, not an absolute pressure. The engine fundamentally responds to air mass, not vacuum, and absolute pressure is necessary to calculate mass. The mass of air entering the engine is directly proportional to the air density, which is proportional to the absolute pressure, and inversely proportional to the absolute temperature.

Carburetors are largely dependent on air volume flow and vacuum, and neither directly infers mass. Consequently, carburetors are precise, but not accurate fuel metering devices. Carburetors were replaced by more accurate fuel metering methods, such as fuel injection in combination with an air mass flow sensor (MAF).

EGR Testing

With OBD II standards, vehicle manufacturers were required to test the exhaust gas recirculation (EGR) valve for functionality during driving. Some manufacturers use the MAP sensor to accomplish this. In these vehicles, they have a MAF sensor for their primary load sensor. The MAP sensor is then used for rationality checks and to test the EGR valve. The way they do this is during a deceleration of the vehicle when there is low absolute pressure in the intake manifold (i.e., a high vacuum present in the intake manifold relative to the outside air) the powertrain control module

(PCM) will open the EGR valve and then monitor the MAP sensor's values. If the EGR is functioning properly, the manifold absolute pressure will increase as exhaust gases enter.

Common Confusion with Boost Sensors and Gauges

MAP sensors measure absolute pressure. Boost sensors or gauges measure the amount of pressure above a set absolute pressure. That set absolute pressure is usually 100 kPa. This is commonly referred to as gauge pressure. Boost pressure is relative to absolute pressure - as one increases or decreases, so does the other. It is a one-to-one relationship with an offset of -100 kPa for boost pressure. Thus a MAP sensor will always read 100 kPa more than a boost sensor measuring the same conditions. A MAP sensor will never display a negative reading because it is measuring absolute pressure, where zero is the total absence of pressure. Vacuum is measured as a negative pressure relative to normal atmospheric pressure. Vacuum-Boost sensors can display negative readings, indicating vacuum or suction (a condition of lower pressure than the surrounding atmosphere). In forced induction engines (supercharged or turbocharged), a negative boost reading indicates that the engine is drawing air faster than it is being supplied, creating suction. The suction is caused by throttling in spark ignition engines and is not present in diesel engines. This is often called vacuum pressure when referring to internal combustion engines.

In short, in a standard atmosphere most boost sensors will read one atmosphere less than a MAP sensor reads. At sea level one can convert boost to MAP by adding approximately 100 kPa. One can convert from MAP to boost by subtracting 100 kPa.

Starter

A starter (also self-starter, cranking motor, or starter motor) is a device used to rotate (crank) an internal-combustion engine so as to initiate the engine's operation under its own power. Starters can be electric, pneumatic, or hydraulic. In the case of very large engines, the starter can even be another internal-combustion engine.

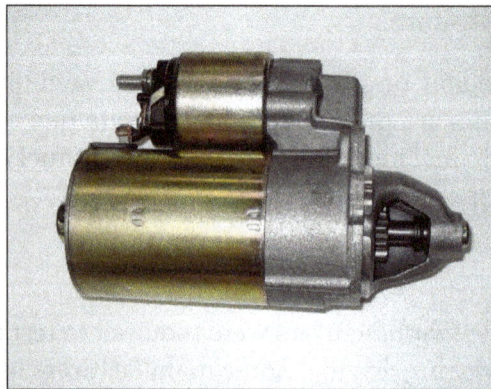

An automobile starter motor (larger cylinder). The smaller object on top
is a starter solenoid which controls power to the starter motor.

Internal combustion engines are feedback systems, which, once started, rely on the inertia from each cycle to initiate the next cycle. In a four-stroke engine, the third stroke releases energy from the fuel, powering the fourth (exhaust) stroke and also the first two (intake, compression) strokes of the next cycle, as well as powering the engine's external load. To start the first cycle at the beginning of any particular session, the first two strokes must be powered in some other way than from the engine itself. The starter motor is used for this purpose and is not required once the engine starts running and its feedback loop becomes self-sustaining.

Starter ring gear on its flywheel.

Electric

The electric starter motor or cranking motor is the most common type used on gasoline engines and small diesel engines. The modern starter motor is either a permanent-magnet or a series-parallel wound direct current electric motor with a starter solenoid (similar to a relay) mounted on it. When DC power from the starting battery is applied to the solenoid, usually through a key-operated switch (the "ignition switch"), the solenoid engages a lever that pushes out the drive pinion on the starter driveshaft and meshes the pinion with the starter ring gear on the flywheel of the engine.

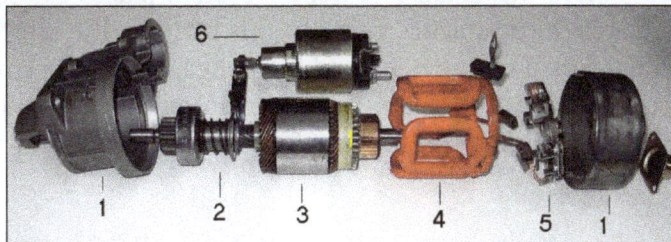

1. Main housing (yoke). 2. Freewheel and pinion gear assembly. 3. Armature.
4. Field coils with brushes attached. 5. Brush-carrier. 6. Solenoid.

The solenoid also closes high-current contacts for the starter motor, which begins to turn. Once the engine starts, the key-operated switch is opened, a spring in the solenoid assembly pulls the pinion gear away from the ring gear, and the starter motor stops. The starter's pinion is clutched to its drive shaft through an overrunning sprag clutch which permits the pinion to transmit drive in only one direction. In this manner, drive is transmitted through the pinion to the flywheel ring gear, but if the pinion remains engaged (as for example because the operator fails to release the key as soon as the engine starts, or if there is a short and the solenoid remains engaged), the pinion will spin independently of its drive shaft. This prevents the engine driving the starter, for such backdrive would cause the starter to spin so fast as to fly apart.

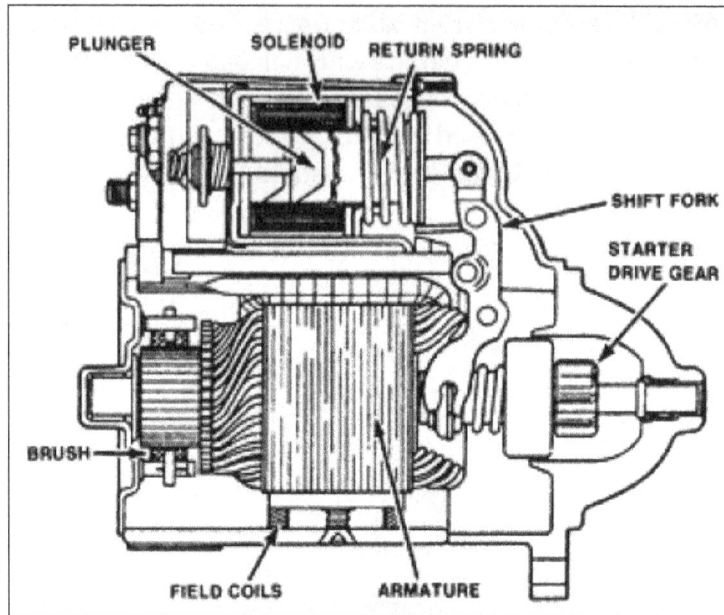

Starter motor diagram.

The sprag clutch arrangement would preclude the use of the starter as a generator if employed in the hybrid scheme mentioned above, unless modifications were made. The standard starter motor is typically designed for intermittent use, which would preclude its use as a generator. The starter's electrical components are designed only to operate for typically under 30 seconds before overheating (by too-slow dissipation of heat from ohmic losses), to save weight and cost. Most automobile owner manuals instruct the operator to pause for at least ten seconds after each ten or fifteen seconds of cranking the engine, when trying to start an engine that does not start immediately.

This overrunning-clutch pinion arrangement was phased into use beginning in the early 1960s; before that time, a Bendix drive was used. The Bendix system places the starter drive pinion on a helically cut drive shaft. When the starter motor begins turning, the inertia of the drive pinion assembly causes it to ride forward on the helix and thus engage with the ring gear. When the engine starts, backdrive from the ring gear causes the drive pinion to exceed the rotative speed of the starter, at which point the drive pinion is forced back down the helical shaft and thus out of mesh with the ring gear.

Folo-thru Drive

An intermediate development between the Bendix drive developed in the 1930s and the overrunning-clutch designs introduced in the 1960s was the Bendix Folo-Thru drive. The standard Bendix drive would disengage from the ring gear as soon as the engine fired, even if it did not continue to run. The Folo-Thru drive contains a latching mechanism and a set of flyweights in the body of the drive unit. When the starter motor begins turning and the drive unit is forced forward on the helical shaft by inertia, it is latched into the engaged position. Only once the drive unit is spun at a speed higher than that attained by the starter motor itself (i.e., it is backdriven by the running engine) will the flyweights pull radially outward, releasing the latch and permitting the overdriven drive unit to be spun out of engagement. In this manner, unwanted starter disengagement is avoided before a successful engine start.

Gear Reduction

In 1962, Chrysler introduced a starter incorporating a geartrain between the motor and the drive shaft. The motor shaft included integrally cut gear teeth forming a pinion that meshes with a larger adjacent driven gear to provide a gear reduction ratio of 3.75:1. This permitted the use of a higher-speed, lower-current, lighter and more compact motor assembly while increasing cranking torque. Variants of this starter design were used on most rear- and four-wheel-drive vehicles produced by Chrysler Corporation from 1962 through 1987. It makes a unique, distinct sound when cranking the engine, which led to it being nicknamed the "Highland Park Hummingbird"—a reference to Chrysler's headquarters in Highland Park, Michigan.

The Chrysler gear-reduction starter formed the conceptual basis for the gear-reduction starters that now predominate in vehicles on the road. Many Japanese automakers phased in gear reduction starters in the 1970s and 1980s. Light aircraft engines also made extensive use of this kind of starter, because its light weight offered an advantage.

Those starters not employing offset gear trains like the Chrysler unit generally employ planetary epicyclic gear trains instead. Direct-drive starters are almost entirely obsolete owing to their larger size, heavier weight and higher current requirements.

Movable Pole Shoe

Ford issued a nonstandard starter, a direct-drive "movable pole shoe" design that provided cost reduction rather than electrical or mechanical benefits. This type of starter eliminated the solenoid, replacing it with a movable pole shoe and a separate starter relay. This starter operates as follows: The driver turns the key, activating the starter switch. A small electric current flows through the solenoid actuated starter relay, closing the contacts and sending large battery current to the starter motor. One of the pole shoes, hinged at the front, linked to the starter drive, and spring-loaded away from its normal operating position, is swung into position by the magnetic field created by electricity flowing through its field coil. This moves the starter drive forward to engage the flywheel ring gear, and simultaneously closes a pair of contacts supplying current to the rest of the starter motor winding. Once the engine starts and the driver releases the starter switch, a spring retracts the pole shoe, which pulls the starter drive out of engagement with the ring gear.

This starter was used on Ford vehicles from 1973 through 1990, when a gear-reduction unit conceptually similar to the Chrysler unit replaced it.

Inertia Starter

A variant on the electric starter motor is the inertia starter. Here, the starter motor does not turn the engine directly. Instead, when energized, the motor turns a heavy flywheel built into its casing (not the main flywheel of the engine). Once the flywheel/motor unit has reached a constant speed the current to the motor is turned off and the drive between the motor and flywheel is disengaged by a freewheel mechanism. The spinning flywheel is then connected to the main engine and its inertia turns it over to start it. These stages are commonly automated by solenoid switches, with the machine operator using a two-position control switch, which is held in one position to spin the motor and then moved to the other to cut the current to the motor and engage the flywheel to the engine.

The advantage of the inertia starter is that, because the motor is not driving the engine directly, it can be of much lower power than the standard starter for an engine of the same size. This allows for a motor of much lower weight and smaller size, as well as lighter cables and smaller batteries to power the motor. This made the inertia starter a common choice for aircraft with large radial piston engines. The disadvantage is the increased time required to start the engine - spinning up the flywheel to the required speed can take between 10 and 20 seconds. If the engine does not start by the time the flywheel has lost its inertia then the process must be repeated for the next attempt.

Pneumatic

Some gas turbine engines and diesel engines, particularly on trucks, use a pneumatic self-starter. In ground vehicles the system consists of a geared turbine, an air compressor and a pressure tank. Compressed air released from the tank is used to spin the turbine, and through a set of reduction gears, engages the ring gear on the flywheel, much like an electric starter. The engine, once running, drives the compressor to recharge the tank.

Aircraft with large gas turbine engines are typically started using a large volume of low-pressure compressed air, supplied from a very small engine referred to as an auxiliary power unit, located elsewhere in the aircraft. Alternately, aircraft gas turbine engines can be rapidly started using a mobile ground-based pneumatic starting engine, referred to as a *start cart* or *air start cart*.

On larger diesel generators found in large shore installations and especially on ships, a pneumatic starting gear is used. The air motor is normally powered by compressed air at pressures of 10–30 bar. The air motor is made up of a center drum about the size of a soup can with four or more slots cut into it to allow for the vanes to be placed radially on the drum to form chambers around the drum. The drum is offset inside a round casing so that the inlet air for starting is admitted at the area where the drum and vanes form a small chamber compared to the others. The compressed air can only expand by rotating the drum, which allows the small chamber to become larger and puts another one of the cambers in the air inlet. The air motor spins much too fast to be used directly on the flywheel of the engine; instead a large gearing reduction, such as a planetary gear, is used to lower the output speed. A Bendix gear is used to engage the flywheel.

Since large trucks typically use air brakes, the system does double duty, supplying compressed air to the brake system. Pneumatic starters have the advantages of delivering high torque, mechanical simplicity and reliability. They eliminate the need for oversized, heavy storage batteries in prime mover electrical systems.

Large Diesel generators and almost all Diesel engines used as the prime mover of ships use compressed air acting directly on the cylinder head. This is not ideal for smaller Diesels, as it provides too much cooling on starting. Also, the cylinder head needs to have enough space to support an extra valve for the air start system. The air start system is conceptually very similar to a distributor in a car. There is an air distributor that is geared to the camshaft of the Diesel engine; on the top of the air distributor is a single lobe similar to what is found on a camshaft. Arranged radially around this lobe are roller tip followers for every cylinder. When the lobe of the air distributor hits one of the followers it will send an air signal that acts upon the back of the air start valve located in the cylinder head, causing it to open. Compressed air is provided from a large reservoir that feeds into a header located along the engine. As soon as the air start valve is opened, the compressed air is admitted and the engine will begin turning. It can be used on two-cycle and four-cycle engines and on reversing engines. On large two-stroke engines less than one revolution of the crankshaft is needed for starting.

Hydraulic

Hydraulic Starter.

Some Diesel engines from 6 to 16 cylinders are started by means of a hydraulic motor. Hydraulic starters and the associated systems provide a sparkless, reliable method of engine starting over a wide temperature range. Typically, hydraulic starters are found in applications such as remote generators, lifeboat propulsion engines, offshore fire pumping engines, and hydraulic fracturing rigs. The system used to support the hydraulic starter includes valves, pumps, filters, a reservoir, and piston accumulators. The operator can manually recharge the hydraulic system; this cannot readily be done with electric starting systems, so hydraulic starting systems are favored in applications wherein emergency starting is a requirement.

With various configurations, Hydraulic starters can be fitted on any engine. Hydraulic starters employ the high efficiency of the axial piston motor concept, which provides high torque at any temperature or environment, and guarantees minimal wear of the engine ring gear and the pinion.

Non-motor

Spring Starter

A spring starter uses potential energy stored in a spring wound up with a crank to start an engine without a battery or alternator. Turning the crank moves the pinion into mesh with the engine's ring gear, then winds up the spring. Pulling the release lever then applies the spring tension to the pinion, turning the ring gear to start the engine. The pinion automatically disengages from the flywheel after operation. Provision is also made to allow the engine to be slowly turned over by hand for engine maintenance. This is achieved by operating the trip lever just after the pinion has engaged with the flywheel. Subsequent turning of the winding handle during this operation will not load the starter. Spring starters can be found in engine-generators, hydraulic power packs, and on lifeboat engines, with the most common application being backup starting system on seagoing vessels.

Spring Starter.

Fuel-starting

Some modern gasoline engines with twelve or more cylinders always have at least one or more pistons at the beginning of its power stroke and are able to start by injecting fuel into that cylinder and igniting it. If the engine is stopped at correct position, the procedure can be applied to engines with fewer cylinders. It is one way of starting an engine of a car with stop-start system.

High Tension Leads

High tension leads or high tension cables or spark plug wires or spark plug cables are the wires that connect a distributor, ignition coil, or magneto to each of the spark plugs in some types of internal combustion engine. "High tension lead" or "cable" is also used for any electrical cable carrying a high voltage in any context. *Tension* in this instance is a synonym for voltage. High tension leads, like many engine components, wear out over time. Each lead contains only one wire, as the current does not return through the same lead, but through the earthed/grounded engine which is connected to the opposite battery terminal (negative terminal on modern engines) high tension may also be referred to as HT.

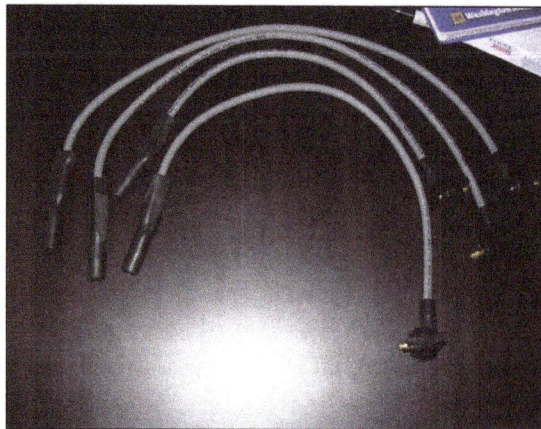
A set of spark plug wires.

Spark plug wires have an outer insulation several times thicker than the conductor, made of a very flexible and heat-resistant material such as silicone or EPDM rubber. The thick insulation prevents arcing from the cable to an earthed engine component. A rubber "boot" covers each terminal. Dielectric grease can be used to improve insulation; a small amount can be applied in the inside of the rubber boot at each end of each wire to help seal out moisture. Printing on spark plug wires may include a brand name, insulation thickness (in millimeters), insulation material type, cylinder number, and conductor type (suppressor or solid wire).

Candle Cable Production.

The wire from each spark plug is just long enough to reach the distributor, without excess. Each end of a spark plug wire has a metal terminal that clips onto the spark plug and distributor, coil, or magneto. There are dedicated spark plug wire pliers, tools designed for removing the terminal from a spark plug without damaging it.

To reduce radio frequency interference (RFI) produced by the spark being radiated by the wires, which may cause malfunction of sensitive electronic systems in modern vehicles or interfere with the car radio, various means in the spark plug and associated lead have been used over time to reduce the nuisance:

- Copper conductors (no suppression).

- Resistor in spark plug with copper conductor.

- Compressed carbon powder as conductor in the lead to act as a resistor.

- Stainless steel wire wound as a coil in the lead with a resistance of about 1300 ohms/meter since 1980s. This acts as an inductor and a resistor.

Application

Placing spark plug wires back into their separators or holders during replacement helps to keep them in place despite engine vibration, extending their life. A common problem with spark plug wires is corrosion of the metal end terminals. Better-quality spark plug wires usually have brass terminals, which are more resistant to corrosion than other metals used.

Older engines also have a wire connecting the ignition coil to the distributor, known as a coil wire. A coil wire is of the same construction as a spark plug wire, but generally shorter and with different terminals. Some distributors have an ignition coil built inside them, eliminating the need for a separate coil wire, e.g. GM High energy ignition system and some Toyotas and Hondas.

Many modern car engines have multiple ignition coils (one for each pair of cylinders) built into a coil pack, eliminating the need for a distributor and coil wire. Some car engines use a small ignition coil mounted on top of each spark plug, eliminating the need for spark plug wires entirely.

Risks and Prevention

The average life span of a candle cable is about 50000 km (in the case of a mixed path). The wear of the spark plug wire leads first to an increase in fuel consumption and a loss in performance. The increased risk of a worn spark plug wire is the possibility of damaging the system downstream of the carburation.

Mass Flow Sensor

A mass air flow sensor (also known as a MAF) is component that is used to meter the amount of air that enters the intake system of an internal combustion engine. Since modern gasoline engines have to maintain tight control over the air/fuel ratio, it is imperative that there be a way to determine how much air is entering the system at any given time, and mass air flow sensors are one way to accomplish this.

There are a number of different technologies that these devices can employ, including spring-loaded door flaps, vortex sensors, membrane sensors, and both hot and cold wire sensors.

Modern internal combustion engines are, in simplest terms, big air pumps. Air enters through the intake, feeds the combustion process, and is expelled through the exhaust. The constant pumping of the pistons, and the powerful vacuum that this action creates, serves to suck in enough air through the intake to enable combustion, and the explosive force of combustion serves to force the air and various byproducts out through the exhaust. This is all rather simple in theory, but modern internal combustion engines are also finely-tuned machines that require *precise* mixture ratios of air and fuel to operate properly.

Since air is a gas, the *mass* of a single *volume* of air can vary widely depending on factors like pressure and temperature. As atmospheric pressure drops, any given volume of air will have less mass than it did at a higher pressure, and the same is true for high temperatures versus low temperatures. That's why simply measuring the volume of air that passes through the intake won't work. In order for an engine control unit (ECU) to "know" how to set the air/fuel mixture, it has to know exactly how much air is entering the system at any given time.

There are a lot of different ways to measure the mass of any given volume of air, both directly and indirectly, but the vast majority of MAFs use one of two different methods: vane meter and hot wire. These are both indirect methods that require additional sensor inputs, which means that if those sensors fail, the ECU isn't able to properly calculate the mass of the air entering the intake.

Vane meter sensors, which are also known as volume air flow (VAF) sensors, are mechanical in nature. They consist of a housing that contains a flap or door that is spring-loaded. The housing is mounted in-between the air filter and the intake manifold, so any air that enters the intake has to pass through it. As the intake draws air in, the flap is forced open. The volume of the air can then be determined by measuring the angle of the flap. To that end, this type of sensor includes a potentiometer that is attached to the flap.

Of course, simply knowing the volume of the air that enters the intake system isn't enough. To that end, this type of MAF often includes a built-in intake air temperature (IAT) sensor. Between the volume of the air and the temperature of the air, the ECU is then able to calculate its mass with some degree of accuracy.

The other main type of mass air flow sensor is known as a "hot wire sensor." This type of MAF is also installed in-line between the air filter and the intake manifold, but it doesn't use a flap. Instead, there is a thin wire suspended inside the housing. This wire is "hot" in that it has a current passing through it. Since the resistance of the wire increases and decreases depending on its temperature, that resistance can be used to calculate roughly how much air is passing through the MAF at any given time.

Unlike vane sensors, hot wire MAFs are able to measure mass without an additional temperature sensor. This is due to the fact that dense air will tend to cool the wire down more than air that is less dense, which allows the resistance reading of the wire to indicate the mass of the air rather than simple its volume.

Other types of mass air flow sensors, including cold wire and vortex sensors, are used less frequently. Cold wire sensors measure the inductance of a sensor that changes depending on how much air is flowing over it, and vortex sensors operate by creating an oscillating series of Karman vortices within the sensor housing. There are a couple of different ways that these vortices can be measured, but in any case it is the frequency of the oscillations that allows the air mass to be determined.

Since, modern engines depend on mass air flow sensors to correctly set the air/fuel mixture, a failed mass air flow sensor can cause a whole slew of driveability issues. You may experience a significant drop in fuel economy, a lack of power, or the engine may not even run well enough to drive the vehicle. There are some cases where a fouled MAF can be repaired and returned to service, but in most cases these (typically expensive) components simply have to be replaced.

References

- Main-parts-of-internal-combustion-engine-or-automobile-engine: mech4study.com, Retrieved 14 July, 2019

- "Racing piston technology – piston weight and design – circle track magazine". Hot rod network. 2007-05-31. Retrieved 2018-04-22

- Hartman, jeff (2007). Turbocharging performance handbook. Motorbooks international. P. 95. Isbn 978-1-61059-231-4

- "Forgotten hero: the man who invented the two-stroke engine". David boothroyd, the vu. Archived from the original on 2004-12-15. Retrieved 2005-01-19

- Mass-air-flow-sensor: crankshift.com, Retrieved 17 May, 2019

4
Fuel and Ignition Systems

The fuel system is made up of fuel filter, fuel pump, carburetor and fuel injector. The injection system generates a spark to ignite a fuel-air mixture within internal combustion engines. The diverse applications of these components of the fuel and ignition system have been thoroughly discussed in this chapter.

Carburetor

Carburetor is a device for supplying a spark-ignition engine with a mixture of fuel and air. Components of carburetors usually include a storage chamber for liquid fuel, a choke, an idling (or slow-running) jet, a main jet, a venturi-shaped air-flow restriction, and an accelerator pump. The quantity of fuel in the storage chamber is controlled by a valve actuated by a float. The choke, a butterfly valve, reduces the intake of air and allows a fuel-rich charge to be drawn into the cylinders when a cold engine is started. As the engine warms up, the choke is gradually opened either by hand or automatically by heat- and engine-speed-responsive controllers. The fuel flows out of the idling jet into the intake air as a result of reduced pressure near the partially closed throttle valve. The main fuel jet comes into action when the throttle valve is further open. Then the venturi-shaped air-flow restriction creates a reduced pressure for drawing fuel from the main jet into the air stream at a rate related to the air flow so that a nearly constant fuel-air ratio is obtained. The accelerator pump injects fuel into the inlet air when the throttle is opened suddenly.

In the 1970s, new legislation and consumer preferences led automobile manufacturers to improve fuel efficiency and lower pollutant emissions. To accomplish these objectives, engineers developed fuel injection management systems based on new computer technologies. Soon, fuel injection systems replaced carbureted fuel systems in virtually all gasoline engines except for two-cycle and small four-cycle gasoline engines, such as those used in lawn mowers.

Fuel Injection

Fuel injection is the introduction of fuel in an internal combustion engine, most commonly automotive engines, by the means of an injector.

All diesel engines use fuel injection by design. Petrol engines can use gasoline direct injection, where the fuel is directly delivered into the combustion chamber, or indirect injection where the fuel is mixed with air before the intake stroke.

On petrol engines, fuel injection replaced carburetors from the 1980s onward. The primary difference between carburetion and fuel injection is that fuel injection atomizes the fuel through a small nozzle under high pressure, while a carburetor relies on suction created by intake air accelerated through a Venturi tube to draw the fuel into the airstream.

Fuel rail connected to the injectors that are mounted just above the intake manifold on a four-cylinder engine.

Objectives

The functional objectives for fuel injection systems can vary. All share the central task of supplying fuel to the combustion process, but it is a design decision how a particular system is optimized. There are several competing objectives such as:

- Power output
- Fuel efficiency
- Emissions performance
- Running on alternative fuels
- Reliability
- Driveability and smooth operation
- Initial cost
- Maintenance cost
- Diagnostic capability
- Range of environmental operation
- Engine tuning

Modern digital electronic fuel injection systems optimize these competing objectives more effectively and consistently than earlier fuel delivery systems (such as carburetors). Carburetors have the potential to atomize fuel better.

Benefits

Benefits of fuel injection include smoother and more consistent transient throttle response, such as during quick throttle transitions, easier cold starting, more accurate adjustment to account for extremes of ambient temperatures and changes in air pressure, more stable idling, decreased maintenance needs, and better fuel efficiency.

Fuel injection also dispenses with the need for a separate mechanical choke, which on carburetor-equipped vehicles must be adjusted as the engine warms up to normal temperature. Furthermore, on spark ignition engines, (direct) fuel injection has the advantage of being able to facilitate stratified combustion which has not been possible with carburetors.

It is only with the advent of multi-point fuel injection certain engine configurations such as in-line five-cylinder gasoline engines have become more feasible for mass production, as traditional carburetor arrangements with single or twin carburetors can not provide even fuel distribution between cylinders, unless a more complicated individual carburetor per cylinder is used.

Fuel injection systems are also able to operate normally regardless of orientation, whereas carburetors with floats are not able to operate upside down or in microgravity, such as that encountered on airplanes.

Environmental Benefits

Fuel injection generally increases engine fuel efficiency. With the improved cylinder-to-cylinder fuel distribution of multi-point fuel injection, less fuel is needed for the same power output (when cylinder-to-cylinder distribution varies significantly, some cylinders receive excess fuel as a side effect of ensuring that all cylinders receive *sufficient* fuel).

Exhaust emissions are cleaner because the more precise and accurate fuel metering reduces the concentration of toxic combustion byproducts leaving the engine. The more consistent and predictable composition of the exhaust makes emissions control devices such as catalytic converters more effective and easier to design.

Elimination of Carburetors

In the 1970s and 1980s in the U.S. and Japan, the respective federal governments imposed increasingly strict exhaust emission regulations. During that time period, the vast majority of gasoline-fueled automobile and light truck engines did not use fuel injection. To comply with the new regulations, automobile manufacturers often made extensive and complex modifications to the engine carburetors. While a simple carburetor system is cheaper to manufacture than a fuel injection system, the more complex carburetor systems installed on many engines in the 1970s were much more costly than the earlier simple carburetors. To more easily comply with emissions regulations, automobile manufacturers began installing fuel injection systems in more gasoline engines during the late 1970s.

Open-loop fuel injection systems had already improved cylinder-to-cylinder fuel distribution and engine operation over a wide temperature range, but did not offer further scope to sufficiently control fuel/air mixtures, in order to further reduce exhaust emissions. Later closed-loop fuel injection systems improved the air–fuel mixture control with an exhaust gas oxygen sensor. Although not part of the injection control, a catalytic converter further reduces exhaust emissions.

Fuel injection was phased in through the latter 1970s and 80s at an accelerating rate, with the German, French, and U.S. markets leading and the UK and Commonwealth markets lagging somewhat. Since the early 1990s, almost all gasoline passenger cars sold in first world markets are equipped with electronic fuel injection (EFI). In Brazil, carburetors were entirely replaced by fuel injection during the 1990s, with the first EFI equipped model built in 1989 (the Volkswagen Gol). The carburetor remains in use in developing countries where vehicle emissions are unregulated and diagnostic and repair infrastructure is sparse. Fuel injection is gradually replacing carburetors in these nations too as they adopt emission regulations conceptually similar to those in force in Europe, Japan, Australia, and North America.

Many motorcycles still use carburetored engines, though all current high-performance designs have switched to EFI.

NASCAR finally replaced carburetors with fuel-injection, starting at the beginning of the 2012 NASCAR Sprint Cup Series season.

System Components

The process of determining the necessary amount of fuel, and its delivery into the engine, are known as fuel metering. Early injection systems used mechanical methods to meter fuel, while nearly all modern systems use electronic metering.

Determining How Much Fuel to Supply

The primary factor used in determining the amount of fuel required by the engine is the amount (by weight) of air that is being taken in by the engine for use in combustion. Modern systems use a mass airflow sensor to send this information to the engine control unit.

Data representing the amount of power output desired by the driver (sometimes known as "engine load") is also used by the engine control unit in calculating the amount of fuel required. A throttle position sensor (TPS) provides this information. Other engine sensors used in EFI systems include a coolant temperature sensor, a camshaft or crankshaft position sensor (some systems get the position information from the distributor), and an oxygen sensor which is installed in the exhaust system so that it can be used to determine how well the fuel has been combusted, therefore allowing closed loop operation.

Supplying the Fuel to the Engine

Fuel is transported from the fuel tank (via fuel lines) and pressurised using fuel pumps. Maintaining the correct fuel pressure is done by a fuel pressure regulator. Often a fuel rail is used to divide the fuel supply into the required number of cylinders. The fuel injector injects liquid fuel into the intake air (the location of the fuel injector varies between systems).

Unlike carburetor-based systems, where the float chamber provides a reservoir, fuel injected systems depend on an uninterrupted flow of fuel. To avoid fuel starvation when subject to lateral G-forces, vehicles are often provided with an anti-surge vessel, usually integrated in the fuel tank, but sometimes as a separate, small anti-surge tank.

EFI Gasoline Engine Components

Animated cut through diagram of a typical fuel injector.

These examples specifically apply to a modern EFI gasoline engine. Parallels to fuels other than gasoline can be made, but only conceptually.

- Injectors,

- Fuel pump,

- Fuel pressure regulator,

- Engine control unit,

- Wiring harness,

- Various sensors:

  - Crank/cam position: Hall effect sensor,

  - Airflow: MAF sensor, sometimes this is inferred with a MAP sensor,

  - Exhaust gas oxygen: oxygen sensor, EGO sensor, UEGO sensor.

Engine Control Unit

The engine control unit is central to an EFI system. The ECU interprets data from input sensors to, among other tasks, calculate the appropriate amount of fuel to inject.

Fuel Injector

When signaled by the engine control unit the fuel injector opens and sprays the pressurised fuel into the engine. The duration that the injector is open (called the pulse width) is proportional to the amount of fuel delivered. Depending on the system design, the timing of when injector opens is either relative each individual cylinder (for a sequential fuel injection (SFI) system), or injectors for multiple cylinders may be signalled to open at the same time (in a batch fire system).

Target Air–fuel Ratios

The relative proportions of air and fuel vary according to the type of fuel used and the performance requirements (i.e. power, fuel economy, or exhaust emissions).

Various Injection Schemes

Single-point Injection

Single-point injection (SPI) uses a single injector at the throttle body (the same location as was used by carburetors).

It was introduced in the 1940s in large aircraft engines (then called the pressure carburetor) and in the 1980s in the automotive world (called Throttle-body Injection by General Motors, Central Fuel Injection by Ford, PGM-CARB by Honda, and EGI by Mazda). Since the fuel passes through the intake runners (like a carburetor system), it is called a "wet manifold system".

The justification for single-point injection was low cost. Many of the carburetor's supporting components - such as the air cleaner, intake manifold, and fuel line routing - could be reused. This postponed the redesign and tooling costs of these components. Single-point injection was used extensively on American-made passenger cars and light trucks during 1980-1995, and in some European cars in the early and mid-1990s.

Continuous Injection

In a continuous injection system, fuel flows at all times from the fuel injectors, but at a variable flow rate. This is in contrast to most fuel injection systems, which provide fuel during short pulses of varying duration, with a constant rate of flow during each pulse. Continuous injection systems can be multi-point or single-point, but not direct.

The most common automotive continuous injection system is Bosch's K-Jetronic, introduced in 1974. K-Jetronic was used for many years between 1974 and the mid-1990s by BMW, Lamborghini, Ferrari, Mercedes-Benz, Volkswagen, Ford, Porsche, Audi, Saab, DeLorean, and Volvo. Chrysler used a continuous fuel injection system on the 1981-1983 Imperial.

In piston aircraft engines, continuous-flow fuel injection is the most common type. In contrast to automotive fuel injection systems, aircraft continuous flow fuel injection is all mechanical, requiring no electricity to operate. Two common types exist: the Bendix RSA system, and the TCM system. The Bendix system is a direct descendant of the pressure carburetor. However, instead of having a discharge valve in the barrel, it uses a *flow divider* mounted on top of the engine, which controls the discharge rate and evenly distributes the fuel to stainless steel injection lines to the intake ports of each cylinder. The TCM system is even more simple. It has no venturi, no pressure chambers, no diaphragms, and no discharge valve. The control unit is fed by a constant-pressure fuel pump. The control unit simply uses a butterfly valve for the air, which is linked by a mechanical linkage to a rotary valve for the fuel. Inside the control unit is another restriction, which controls the fuel mixture. The pressure drop across the restrictions in the control unit controls the amount of fuel flow, so that fuel flow is directly proportional to the pressure at the flow divider. In fact, most aircraft that use the TCM fuel injection system feature a fuel flow gauge that is actually a pressure gauge calibrated in *gallons per hour* or *pounds per hour* of fuel.

Central Port Injection

From 1992 to 1996, General Motors implemented a system called Central Port Injection or Central Port Fuel Injection. The system uses tubes with poppet valves from a central injector to spray fuel at each intake port rather than the central throttle-body. Fuel pressure is similar to a single-point injection system. CPFI is a batch-fire system, while CSFI (from 1996) is a sequential system.

Multipoint Fuel Injection

Multipoint fuel injection (MPI), also called port fuel injection (PFI), injects fuel into the intake ports just upstream of each cylinder's intake valve, rather than at a central point within an intake manifold. MPI systems can be *sequential*, in which injection is timed to coincide with each cylinder's intake stroke; *batched*, in which fuel is injected to the cylinders in groups, without precise synchronization to any particular cylinder's intake stroke; or *simultaneous*, in which fuel is injected at the same time to all the cylinders. The intake is only slightly wet, and typical fuel pressure runs between 40-60 psi.

Many modern EFI systems use sequential MPI; however, in newer gasoline engines, direct injection systems are beginning to replace sequential ones.

Direct Injection

In a direct injection engine, fuel is injected into the combustion chamber as opposed to injection before the intake valve (petrol engine) or a separate pre-combustion chamber (diesel engine).

In a common rail system, the fuel from the fuel tank is supplied to the common header (called the accumulator). This fuel is then sent through tubing to the injectors, which inject it into the combustion chamber. The header has a high pressure relief valve to maintain the pressure in the header and return the excess fuel to the fuel tank. The fuel is sprayed with the help of a nozzle that is opened and closed with a needle valve, operated with a solenoid. When the solenoid is not activated, the spring forces the needle valve into the nozzle passage and prevents the injection of fuel into the cylinder. The solenoid lifts the needle valve from the valve seat, and fuel under pressure is sent in the engine cylinder. Third-generation common rail diesels use piezoelectric injectors for increased precision, with fuel pressures up to 1,800 bar or 26,000 psi.

Direct fuel injection costs more than indirect injection systems: the injectors are exposed to more heat and pressure, so more costly materials and higher-precision electronic management systems are required.

Diesel Engines

All Diesel engines have fuel injected into the combustion chamber. Earlier systems, relying on simpler injectors, often injected into a sub-chamber shaped to swirl the compressed air and improve combustion; this was known as indirect injection. However, this was less efficient than the now common direct injection in which initiation of combustion takes place in a depression (often toroidal) in the crown of the piston.

Throughout the early history of diesels, they were always fed by a mechanical pump with a small separate chamber for each cylinder, feeding separate fuel lines and individual injectors. Most such pumps were in-line, though some were rotary.

Most modern diesel engines use common rail or unit injector direct injection systems. A special type of direct injection system is the M-System, that was used throughout the second half of the 20th century.

Gasoline Engines

Modern gasoline engines also use direct injection, which is referred to as gasoline direct injection. This is the next step in evolution from multi-point fuel injection, and offers another magnitude of emission control by eliminating the "wet" portion of the induction system along the inlet tract.

By virtue of better dispersion and homogeneity of the directly injected fuel, the cylinder and piston are cooled, thereby permitting higher compression ratios and earlier ignition timing, with resultant enhanced power output. More precise management of the fuel injection event also enables better control of emissions. Finally, the homogeneity of the fuel mixture allows for leaner air–fuel ratios, which together with more precise ignition timing can improve fuel efficiency. Along with this, the engine can operate with stratified (lean-burn) mixtures, and hence avoid throttling losses at low and part engine load. Some direct-injection systems incorporate piezoelectronic fuel injectors. With their extremely fast response time, multiple injection events can occur during each cycle of each cylinder of the engine.

Swirl Injection

Swirl injectors are used in liquid rocket, gas turbine, and diesel engines to improve atomization and mixing efficiency.

The circumferential velocity component is first generated as the propellant enters through helical or tangential inlets producing a thin, swirling liquid sheet. A gas-filled hollow core is then formed along the centerline inside the injector due to centrifugal force of the liquid sheet. Because of the presence of the gas core, the discharge coefficient is generally low. In swirl injector, the spray cone angle is controlled by the ratio of the circumferential velocity to the axial velocity and is generally wide compared with nonswirl injectors.

Fuel Pump

A fuel pump is a frequently (but not always) essential component on a car or other internal combustion engined device. Many engines (older motorcycle engines in particular) do not require any fuel pump at all, requiring only gravity to feed fuel from the fuel tank or under high pressure to the fuel injection system. Often, carbureted engines use low pressure mechanical pumps that are mounted outside the fuel tank, whereas fuel injected engines often use electric fuel pumps that are mounted inside the fuel tank (and some fuel injected engines have two fuel pumps: one low pressure/high volume supply pump in the tank and one high pressure/low volume pump on or near the engine). Fuel pressure needs to be within certain specifications for the engine to run correctly. If the fuel pressure is too high, the engine will run rough and rich, not combusting all of the fuel being pumped making the engine inefficient and a pollutant. If the pressure is too low, the engine may run lean, misfire, or stall.

A high-pressure fuel pump on a Yanmar 2GM20 marine diesel engine.

Mechanical Pump

Prior to the widespread adoption of electronic fuel injection, most carbureted automobile engines used mechanical fuel pumps to transfer fuel from the fuel tank into the fuel bowls of the carburetor. The two most widely used fuel feed pumps are diaphragm and plunger-type mechanical pumps. Diaphragm pumps are a type of positive displacement pump. Diaphragm pumps contain a pump chamber whose volume is increased or decreased by the flexing of a flexible diaphragm, similar to the action of a piston pump. A check valve is located at both the inlet and outlet ports of the pump chamber to force the fuel to flow in one direction only. Specific designs vary, but in the most common configuration, these pumps are typically bolted onto the engine block or head, and the engine's camshaft has an extra eccentric lobe that operates a lever on the pump, either directly or via a pushrod, by pulling the diaphragm to bottom dead center. In doing so, the volume inside the pump chamber increased, causing pressure to decrease. This allows fuel to be pushed into the pump from the tank (caused by atmospheric pressure acting on the fuel in the tank). The return motion of the diaphragm to top dead center is accomplished by a diaphragm spring, during which the fuel in the pump chamber is squeezed through the outlet port and into the carburetor. The pressure at which the fuel is expelled from the pump is thus limited (and therefore regulated) by the force applied by the diaphragm spring.

The carburetor typically contains a float bowl into which the expelled fuel is pumped. When the fuel level in the float bowl exceeds a certain level, the inlet valve to the carburetor will close, preventing the fuel pump from pumping more fuel into the carburetor. At this point, any remaining fuel inside the pump chamber is trapped, unable to exit through the inlet port or outlet port. The diaphragm will continue to allow pressure to the diaphragm, and during the subsequent rotation, the eccentric will pull the diaphragm back to bottom dead center, where it will remain until the inlet valve to the carburetor reopens.

Mechanical fuel pump, fitted to cylinder head.

Because one side of the pump diaphragm contains fuel under pressure and the other side is connected to the crankcase of the engine, if the diaphragm splits (a common failure), it can leak fuel into the crankcase. The capacity of both mechanical and electric fuel pump is measured in psi (which stands for pounds per square inch). Usually, this unit is the general measurement for pressure, yet it has slightly different meaning, when talking about fuel pumps.

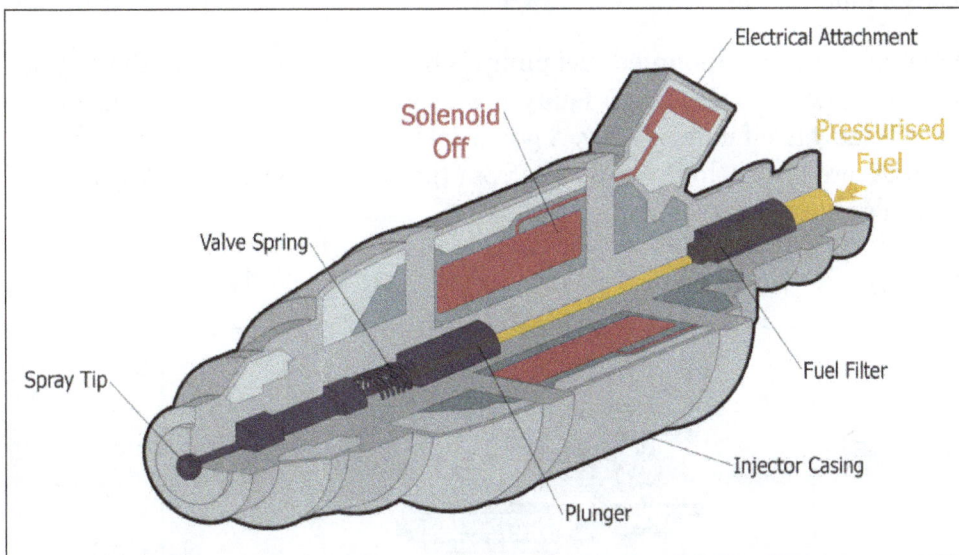

Diagram of diaphragm type fuel pump.

Plunger-type Fuel Pump

Plunger-type pumps are a type of positive displacement pump that contain a pump chamber whose volume is increased and/or decreased by a plunger moving in and out of a chamber full of fuel with inlet and discharge stop-check valves. It is similar to that of a piston pump, but the high-pressure seal is stationary while the smooth cylindrical plunger slides through the seal. These pumps typically run at a higher pressure than diaphragm type pumps. Specific designs vary, but in the most common configuration, these pumps are mounted on the side of the injection pump and driven by the camshaft, either directly or via a pushrod. When the camshaft lobe is at top dead center, the plunger has just finished pushing the fuel through the discharge valve. A spring is used to pull the plunger outward creating a lower pressure pulling fuel into the chamber from the inlet valve. These pumps can run between 250 and 1,800 bar (3,625 and 26,000 psi). Because it is connected to the camshaft, the discharge pressure of these pumps is constant, but the rate at which it pumps is directly correlated to the revolutions per minute (rpm) of the engine.

Both pumps create negative pressure to draw the fuel through the lines. However, the low pressure between the pump and the fuel tank, in combination with heat from the engine and/or hot weather, can cause the fuel to vaporize in the supply line. This results in fuel starvation as the fuel pump, designed to pump liquid, not vapor, is unable to suck more fuel to the engine, causing the engine to stall. This condition is different from vapor lock, where high engine heat on the pressured side of the pump (between the pump and the carburetor) boils the fuel in the lines, also starving the engine of enough fuel to run. Mechanical automotive fuel pumps generally do not generate much more than 10–15 psi, which is more than enough for most carburetors.

Decline of Mechanical Pumps

As engines moved away from carburetors and towards fuel injection, mechanical fuel pumps were replaced with electric fuel pumps, because fuel injection systems operate more efficiently at higher fuel pressures (40–60 psi) than mechanical diaphragm pumps can generate. Electric fuel pumps are generally located in the fuel tank, in order to use the fuel in the tank to cool the pump and to ensure a steady supply of fuel.

Another benefit of an in-tank mounted fuel pump is that a suction pump at the engine could suck in air through a (difficult to diagnose) faulty hose connection, while a leaking connection in a pressure line will show itself immediately. A potential hazard of a tank-mounted fuel pump is that all of the fuel lines are under (high) pressure, from the tank to the engine. Any leak will be easily detected, but is also hazardous.

Electric Pump

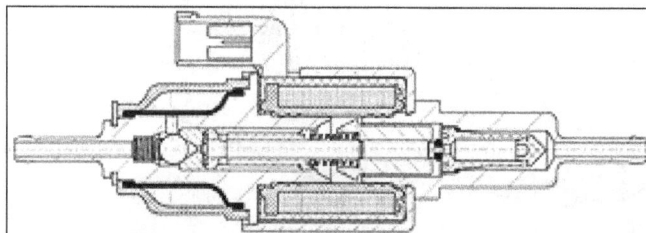

A piston metering pump f.e. gasoline – or additive metering pump.

In many modern cars the fuel pump is usually electric and located inside the fuel tank. The pump creates positive pressure in the fuel lines, pushing the gasoline to the engine. The higher gasoline pressure raises the boiling point. Placing the pump in the tank puts the component least likely to handle gasoline vapor well (the pump itself) farthest from the engine, submersed in cool liquid. Another benefit to placing the pump inside the tank is that it is less likely to start a fire. Though electrical components (such as a fuel pump) can spark and ignite fuel vapors, liquid fuel will not explode and therefore submerging the pump in the tank is one of the safest places to put it. In most cars, the fuel pump delivers a constant flow of gasoline to the engine; fuel not used is returned to the tank. This further reduces the chance of the fuel boiling, since it is never kept close to the hot engine for too long.

Electric fuel pump.

An advantage of an electric fuel pump is reduced fuel consumption because it does not have the resistance associated with a mechanical drive and because the fuel supply can be monitored more accurately by the electronic control unit (ECU). Pre-delivery of fuel can also be accomplished by an electric fuel pump because it does not depend on engine rpm. Due to this, rapid engine starting can be implemented to conserve gas. This is particularly important in stop-start systems where the engine turns itself off when it senses no use, such as stopped at a stoplight.

The ignition switch does not carry the power to the fuel pump; instead, it activates a relay which will handle the higher current load. It is common for the fuel pump relay to become oxidized and cease functioning; this is much more common than the actual fuel pump failing. Modern engines utilize solid-state control which allows the fuel pressure to be controlled via pulse-width modulation of the pump voltage. This increases the life of the pump, allows a smaller and lighter device to be used, and reduces electrical load.

Cars with electronic fuel injection have an electronic control unit (ECU) and this may be programmed with safety logic that will shut the electric fuel pump off, even if the engine is running. In the event of a collision this will prevent fuel leaking from any ruptured fuel line. Additionally, cars may have an inertia switch (usually located underneath the front passenger seat) that is "tripped" in the event of an impact, or a *roll-over valve* that will shut off the fuel pump in case the car rolls over.

Some ECUs may also be programmed to shut off the fuel pump if they detect low or zero oil pressure, for instance if the engine has suffered a terminal failure (with the subsequent risk of fire in the engine compartment).

The fuel sending unit assembly may be a combination of the electric fuel pump, the filter, the strainer, and the electronic device used to measure the amount of fuel in the tank via a float attached to a sensor which sends data to the dash-mounted fuel gauge. The fuel pump by itself is a relatively inexpensive part. But a mechanic at a garage might have a preference to install the entire unit assembly.

High-pressure Fuel Pumps

Pumps for direct-injection engines - such as diesel engines and direct-injection gasoline engines - operate at a much higher pressure due to the engine's specs and are usually mechanical pumps such as common rail radial piston pump, common rail two piston radial, inline pump, port and helix, and metering unit design. Radial piston pumps used in cars and trucks are fuel lubricated, this helps with keeping oil out of the high pressure fuel. Oil in the fuel can raise emissions and clog injectors. Many diesel engines are common rail, which means all the injectors are supplied by one pipe full of high pressure fuel supplied by the fuel pump. Common rails dampen the pressure fluctuations when fuel is used by each injector and pumped into the cylinder. Common rails also make it easier for the fuel pressure to be measured by one device instead of one per cylinder. Port and Helix (plunger type) high pressure fuel pumps are most commonly used in marine diesels because of their simplicity, reliability, and its ability to be scaled up in proportion to the engine size.

Port and Helix Type Pump

Port and Helix pumps are cam driven plunger type pumps that run at one-half engine rpm for four stoke engines and at the same rpm in the case of a two stroke. The pump is similar to that of a radial piston type pump but instead of a piston it has a machined plunger that has no seals. When the plunger is at top dead center, the injection to the cylinder is finished and it is returned on its downward stroke by a compression spring. Because the height of the cam lobe cannot be easily changed, the amount of fuel being pumped to the injector is controlled by a rack and pinion device that rotates the plunger allowing variable amounts of fuel to the area above the plunger. Inlet and outlet ports are located on two sides of the pumps cylinder walls allowing fuel to flow through the compression chamber until the plunger is driven up closing the two ports and starting the compression motion. The outlet port feeds back into the fuel tank/settler of the engine. The fuel is then forced through a stop check valve, to prevent backflow to the injector nozzle at pressures that can exceed 18,000 psi.

Turbopumps

Many jet engines, including rocket engines use a turbopump which is a centrifugal pump usually propelled by a gas turbine or in some cases a ram-air device (particularly in ramjet engines which lack a shaft).

Fuel Filter

A filter cleans the air, is known as air filter, fuel filter cleans the fuel. Lubrication oil is filtered by lubrication filter.

Dirt is dangerous for engine parts. It can be entered into the engine during suction stroke along with fuel and lube oil. Hence to get rid of this unwanted dirt, filters are used.

The diesel engine life is mostly depended on the quality of fuel and lubrication of internal parts. Unfiltered fuel may contain several kinds of contamination. These substances must be removed before the fuel enters the system.

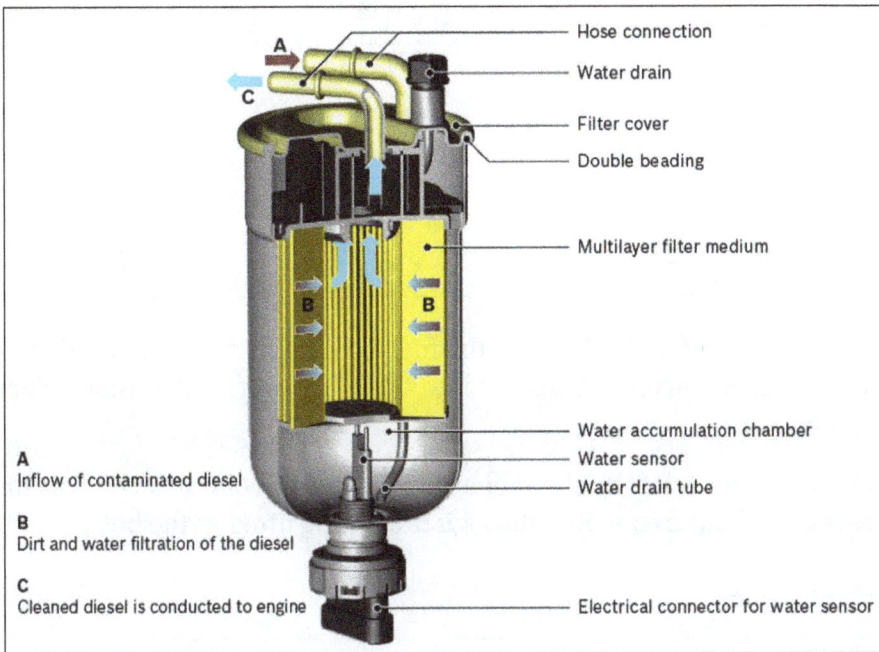

A
Inflow of contaminated diesel

B
Dirt and water filtration of the diesel

C
Cleaned diesel is conducted to engine

Hose connection
Water drain
Filter cover
Double beading
Multilayer filter medium
Water accumulation chamber
Water sensor
Water drain tube
Electrical connector for water sensor

The dirt and foreign material cause rapid wear and failure of the fuel pump, and injector. Also many other Internal Combustion Engine Parts.

Fuel filters serve a vital function in today's modern and compact engine fuel system. It also improves performance, as the fewer contaminants present in the fuel.

If a filter is not replaced regularly it may become clogged with contaminants. It will cause a restriction in the fuel flow. So, it will cause an appreciable drop in engine performance.

Various types of fuel filters are available to make the fuel oil free from dirt and other foreign material. They will suit varying conditions of operations. They included paper element, cloth element, felt element and also the combination of cloth and felt element. There are two types of fuel filter:

Preliminary Fuel Filter

This type of filter is fitted in between fuel tank and fuel feed pump. This filter consists of a single bowl, in which a perforated tube is fitted centrally.

The perforated centre tube is surrounded annularly by a filter element. A gasket is placed on top of the bowl, to check for any leakage of fuel.

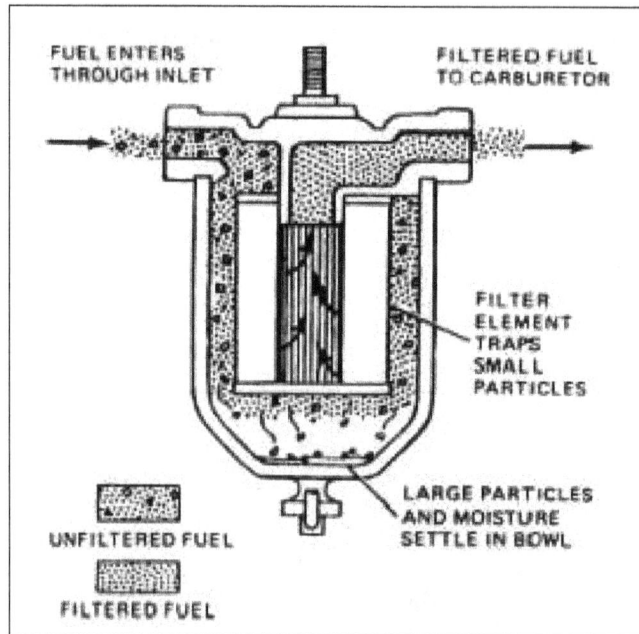

When suction created by the fuel feed pump, the fuel from a fuel tank is sucked through the filter element. It will drop the impurities and enter into the centre perforated tube. Where it is drawn out by the feed pump.

The dirt and other impurities left by the fuel at filter element collect at the bottom of the bowl. Where these are removed out frequently, through drain plug fitted at the bottom of the bowl.

Secondary Fuel Filter

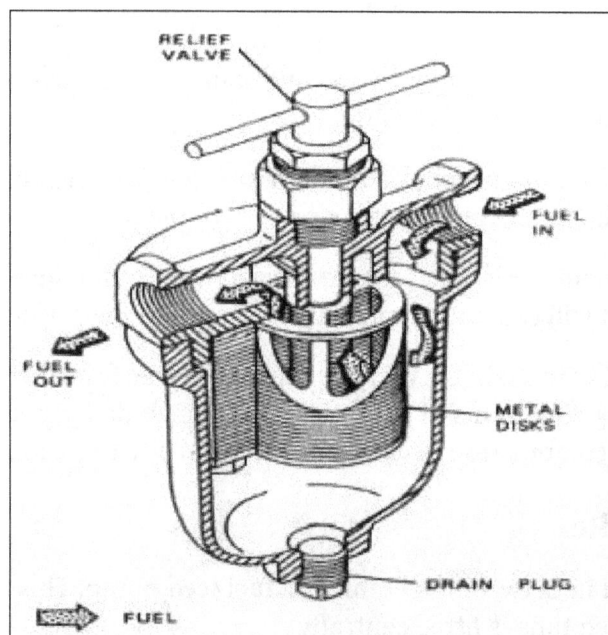

In dual (Two Stage) secondary filter two numbers of filter elements are employed. One having ordinary filter element and other having fine filter element. Each filter unit is supported by the central bolt, which is known as filter element carrier.

The fuel enters the first filter unit (Ordinary Filter Element). Where it drops large dust particles at the element, and then this fuel flows out centrally.

This filtered fuel enters the next stage (Fine Filter Element). Where even the finest particles of dirt etc. are separated from it. Then this purified fuel flows to the feed pump.

An air vent screw is provided at the top of the filter for air venting. Also, at the bottom of the bowl, drain plugs are provided to remove the impurities, which collect from fuel.

Ignition System

The ignition system is one of the most important systems used in the I.C engines. The spark-ignition engine requires some device to ignite the compressed air-fuel mixture. The ignition takes place inside the cylinder at the end of the compression stroke. Ignition system serves this purpose.

It is a part of the electrical system which carries the electrical current to a current plug. It gives the spark to ignite the air-fuel mixture at the correct time.

Types of Ignition System

There are three types of ignition system used in modern-day vehicles:

- Battery ignition system (or coil ignition system).

- Magneto ignition system.

- Electronic Ignition System.

Both the ignition system is based on the principle of common electromagnetic induction. The battery ignition system is mostly used in passenger cars and light trucks.

In the battery ignition system, the current in the primary winding is supplied by the battery. In the magneto to the ignition system, the magneto produces and supplies the current in the primary winding.

Parts of an Ignition System

- Battery

- Switch ignition distributor

- Ignition coil

- Spark plugs and

- Necessary wiring

Some system uses transistors to reduce the load on the distributor contact points. Other systems use a combination of transistors and magnetic pickup in the distributor.

Compression ignition engine does not have such an ignition system. In a compression ignition engine, only air is compressed in the cylinder. And at the end of the compression stroke, the fuel is injected which catch fire due to the high temperature and pressure of the compressed air.

An Ignition System in the Vehicle

The ignition system supplied high voltage surges of current (as high as 30,000 to volts) the spark plug. These surges produce the electric sparks at the spark plug gap. Spark ignite to set fire to the compressed air-fuel mixture in the combustion chamber.

The sparking must take place at the correct time at the end of the compression stroke in every cycle of operation. At high speed or during part throttle operation, the spark is advanced. So that it occurs somewhat earlier in the cycle, the mixture thus has time to burn and deliver its power.

The ignition system should function efficiently at the high and low speeds of the engine. It should be simple to maintain, light and compact. It should not cause any interference.

Battery Ignition System

The figure shows the battery ignition system for a 4 cylinder engine. A battery of 12 volts is generally employed. There are two basic circuits in the system primary and secondary circuits.

The first circuit has the battery, primary winding of the ignition coil, condenser, and the contact breaker from the primary circuit. Whereas the secondary winding of the ignition coil, distributor and the spark plugs forms the secondary circuits.

The value of the voltage depends upon the number of turns in each coil. The high voltage 10,000 to 20,000 volts then passes to a distributor.

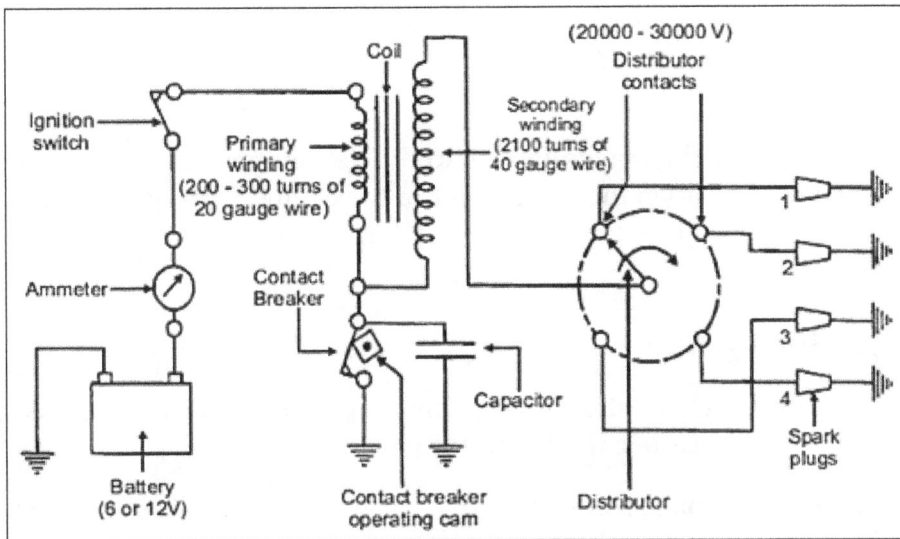

Battery Ignition System.

It consists of the spark plug of the cylinder in rotation depending upon the firing order of the engine. This causing a high-intensity spark jumps across the gap. Thereby, ignition of the air-fuel mixture takes place in all the cylinders. The battery ignition system has massive use in cars, light trucks, buses etc.

Magneto Ignition System

It has the same principle of working like that of the battery ignition system. In this, no battery is required, as the magneto acts as its own generator.

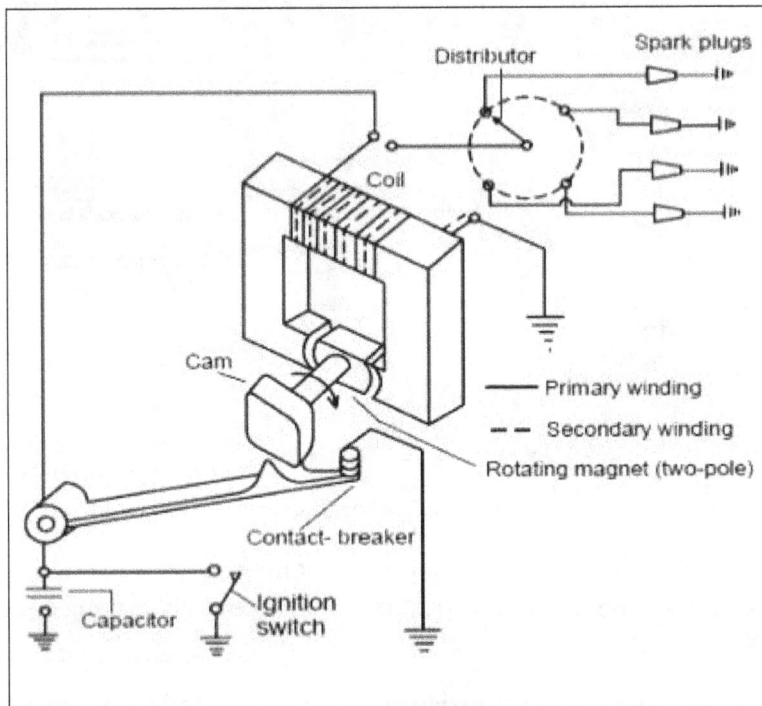

Magneto Ignition System.

It consists of either rotating magnets in fixed coils, or rotating coils in fixed magnets. The current produced by the magneto is made to flow to the induction coil which works in the same as that of the battery ignition system.

This high voltage current is then made to flow to the distributor which connects the sparking plugs in rotation depending upon the firing order of the engine. This type of ignition system is used small spark-ignition engines for example Scooters, Motorcycles and small motorboat engines.

Electronic Ignition System

The conventional electro-mechanical ignition system uses mechanical contact breakers. Though it is very simple, it suffers from certain limitation as follows:

- The contact breaker points handle the heavy current. This resulting in burn out of contact points. Thus it requires periodical servicing and settings.

- The mechanically operated contact breaker has inertial effects. Hence at higher speeds, the make or break of contact may not be timed.

- At higher speeds, the dwell time for building up the current in the coil to its maximum value is low. Thus the spark strength may be reduced.

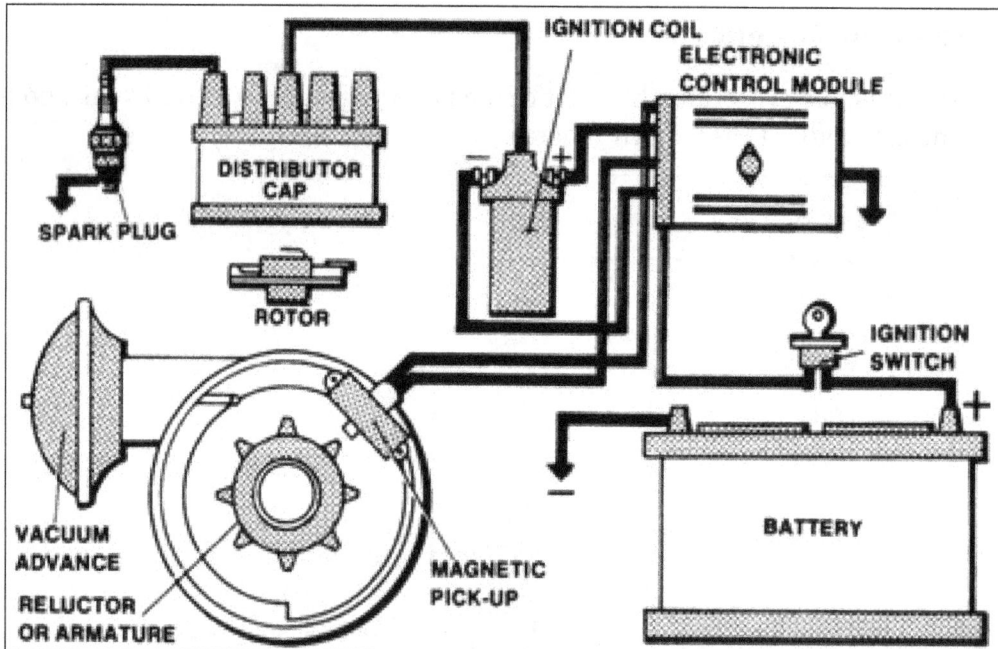

Electronic Ignition System.

To overcome the above drawbacks, in the modern automobiles, electronic ignition systems are used. This electronic ignition system has its best performance at all varying conditions and speed, unlike electro-mechanical systems.

The electro ignition system consists of transistors, capacitors, diodes and resistors. These acts as heavy duty switches in controlling the primary current for the high voltage ignition coil.

Distributor

A distributor is an enclosed rotating shaft used in spark-ignition internal combustion engines that have mechanically-timed ignition. The distributor's main function is to route secondary, or high voltage, current from the ignition coil to the spark plugs in the correct firing order, and for the correct amount of time. Except in magneto systems, the distributor also houses a mechanical or inductive breaker switch to open and close the ignition coil's primary circuit.

The first reliable battery operated ignition was developed by Dayton Engineering Laboratories Co. (Delco) and introduced in the 1910 Cadillac. This ignition was developed by Charles Kettering and was considered a wonder in its day. Atwater Kent invented his Unisparker ignition system about this time in competition with the Delco system. By the end of the 20th century mechanical ignitions were disappearing from automotive applications in favor of inductive or capacitive electronic ignitions fully controlled by engine control units (ECU), rather than directly timed to the engine's crankshaft speed.

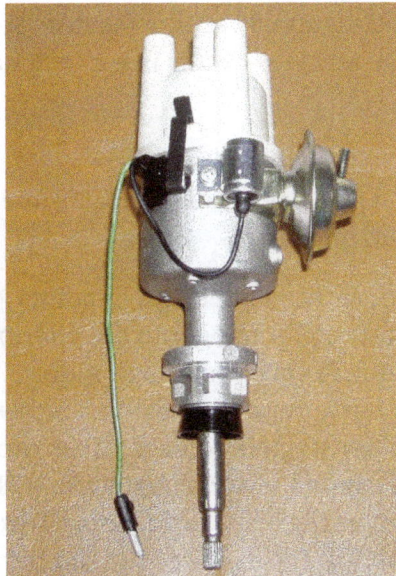

Typical distributor with distributor cap, Also visible are mounting/drive shaft (bottom), vacuum advance unit (right) and capacitor (centre).

A distributor consists of a rotating arm or rotor inside the distributor cap, on top of the distributor shaft, but insulated from it and the body of the vehicle (ground). The distributor shaft is driven by a gear on the camshaft on most overhead valve engines, and attached directly to a camshaft on most overhead cam engines. (The distributor shaft may also drive the oil pump.) The metal part of the rotor contacts the high voltage cable from the ignition coil via a spring-loaded carbon brush on the underside of the distributor cap. The metal part of the rotor arm passes close to (but does not touch) the output contacts which connect via high tension leads to the spark plug of each cylinder. As the rotor spins within the distributor, electric current is able to jump the small gaps created between the rotor arm and the contacts due to the high voltage created by the ignition coil.

The distributor shaft has a cam that operates the contact breaker (also called *points*). Opening the points causes a high induction voltage in the system's ignition coil.

The distributor also houses the centrifugal advance unit: a set of hinged weights attached to the distributor shaft, that cause the breaker points mounting plate to slightly rotate and advance the spark timing with higher engine revolutions per minute (rpm). In addition, the distributor has a vacuum advance unit that advances the timing even further as a function of the vacuum in the inlet manifold. Usually there is also a capacitor attached to the distributor. The capacitor is connected parallel to the breaker points, to suppress sparking to prevent excessive wear of the points.

Around the 1970s, the primary breaker points were largely replaced with a Hall effect sensor or optical sensor. As this is a non-contacting device and the ignition coil is controlled by solid state electronics, a great amount of maintenance in point adjustment and replacement was eliminated. This also eliminates any problem with breaker follower or cam wear, and by eliminating a side load it extends distributor shaft bearing life. The remaining secondary (high voltage) circuit stayed essentially the same, using an ignition coil and a rotary distributor.

Most distributors used on electronically fuel injected engines lack vacuum and centrifugal advance units. On such distributors, the timing advance is controlled electronically by the engine computer. This allows more accurate control of ignition timing, as well as the ability to alter timing based on factors other than engine speed and manifold vacuum (such as engine temperature). Additionally, eliminating vacuum and centrifugal advance results in a simpler and more reliable distributor.

Distributor Cap

The distributor cap is the cover that protects the distributor's internal parts and holds the contacts between internal rotor and the spark plug wires.

The distributor cap has one post for each cylinder, and in points ignition systems there is a central post for the current from the ignition coil coming into the distributor. There are some exceptions however, as some engines (many Alfa Romeo cars, some 1980s Nissans) have *two* spark plugs per cylinder, so there are two leads coming out of the distributor per cylinder. Another implementation is the wasted spark system, where a single contact serves two leads, but in that case each lead connects one cylinder. In General Motors high energy ignition (HEI) systems there is no central post and the ignition coil sits on top of the distributor. Some Toyota and Honda engines also have their coil within the distributor cap. On the inside of the cap there is a terminal that corresponds to each post, and the plug terminals are arranged around the circumference of the cap according to the firing order in order to send the secondary voltage to the proper spark plug at the right time.

The rotor is attached to the top of the distributor shaft which is driven by the engine's camshaft and thus synchronized to it. Synchronization to the camshaft is required as the rotor must turn at exactly half the speed of the main crankshaft in the 4-stroke cycle. Often, the rotor and distributor are attached directly to the end of the one of (or the only) camshaft, at the opposite end to the timing drive belt. This rotor is pressed against a carbon brush on the center terminal of the distributor cap which connects to the ignition coil. The rotor is constructed such that the center tab is electrically connected to its outer edge so the current coming in to the center post travels through the carbon point to the outer edge of the rotor. As the camshaft rotates, the rotor spins and its outer edge passes each of the internal plug terminals to fire each spark plug in sequence.

Engines that use a mechanical distributor may fail if they run into deep puddles because any water that gets onto the distributor can short out the electric current that should go through the spark plugs, rerouting it directly to the body of the vehicle. This in turn causes the engine to stop as the fuel is not ignited in the cylinders. This problem can be fixed by removing the distributor's cap and drying the cap, cam, rotor and the contacts by wiping with tissue paper or a clean rag, by blowing hot air on them, or using a moisture displacement spray e.g. WD-40 or similar. Oil, dirt or other contaminants can cause similar problems, so the distributor should be kept clean inside and outside to ensure reliable operation. Some engines include a rubber o-ring or gasket between the distributor base and cap to help prevent this problem. The gasket is made of a material like Viton or butyl for a tight seal in extreme temperatures and chemical environments. This gasket should not be discarded when replacing the cap. Most distributor caps have the position of the number 1 cylinder's terminal molded into the plastic. By referencing a firing order diagram and knowing the direction the rotor turns, (which can be seen by cranking the engine with the cap off) the spark plug wires can be correctly routed. Most distributor caps are designed so that they cannot be installed in the wrong position. Some older engine designs allow the cap to be installed in the wrong position by 180 degrees, however. The number 1 cylinder position on the cap should be noted before a cap is replaced.

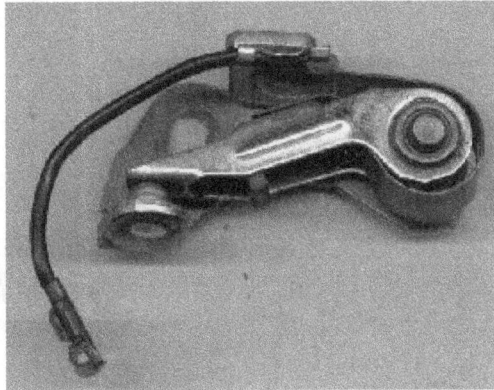

Breaker arm with contact points at the left. The pivot is on the right
and the cam follower is in the middle of the breaker arm.

The distributor cap is a prime example of a component that eventually succumbs to heat and vibration. It is a relatively easy and inexpensive part to replace if its bakelite housing does not break or crack first. Carbon deposit accumulation or erosion of its metal terminals may also cause distributor-cap failure.

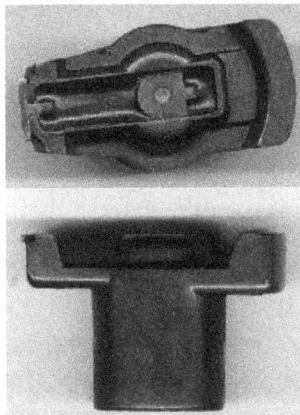

Rotor. This rotates at the same speed as the camshaft, one half the speed of the crankshaft.

As it is generally easy to remove and carry off, the distributor cap can be taken off as a means of theft prevention. Although not practical for everyday use, because it is essential for the starting and running of the engine, its removal thwarts any attempt at hot-wiring the vehicle.

Top of distributor with wires and terminals.   Rotor contacts inside distributor cap.

Direct and Distributorless Ignition

Modern engine designs have abandoned the high-voltage distributor and coil, instead performing the distribution function in the primary circuit electronically and applying the primary (low-voltage) pulse to individual coils for each spark plug, or one coil for each pair of companion cylinders in an engine (two coils for a four-cylinder, three coils for a six-cylinder, four coils for an eight-cylinder, and so on).

In traditional remote distributorless systems, the coils are mounted together in a transformer oil filled 'coil pack', or separate coils for each cylinder, which are secured in a specified place in the engine compartment with wires to the spark plugs, similar to a distributor setup. General Motors, Ford, Chrysler, Hyundai, Subaru, Volkswagen and Toyota are among the automobile manufacturers known to have used coil packs. Coil packs by Delco for use with General Motors engines allow removal of the individual coils in case one should fail, but in most other remote distributorless coil pack setups, if a coil were to fail, replacement of the whole pack would be required to fix the problem.

More recent layouts utilize a coil located very near to (*Coil-Near-Plugs*) or directly on top of each spark plug (Direct Ignition, *DI, coil-on-plug*, or *COP*). This design avoids the need to transmit very high voltages, which is often a source of trouble, especially in damp conditions.

Both direct and remote distributorless systems also allow finer levels of ignition control by the engine computer, which helps to increase power output, decrease fuel consumption and emissions, and implement features such as cylinder deactivation. Spark plug wires, which need routine replacement due to wear, are also eliminated when the individual coils are mounted directly on top of each plug, since the power is transported a very short distance from the coil to the plug.

Four-stroke two-cylinder engines can be built without a distributor, as in the Citroen 2CV of 1948 and BMW boxer twin motorcycles, and some Honda motorcycles from the 1960s (e.g. the CL160 Scrambler). Both spark plugs of the boxer twin are fired simultaneously, resulting in a wasted spark on the cylinder currently on its exhaust stroke.

Four-stroke four-cylinder engines can be built without a distributor, as in the Citroen ID19. Two coils are used with one coil firing two of the spark plugs simultaneously, resulting in a wasted spark on the cylinder currently on its exhaust stroke, and the other coil used for the other two cylinders. This system has been scaled up to engines with virtually an unlimited number of cylinders.

Four-stroke one-cylinder engines can be built without a distributor, as in many lawn mowers, and a growing number of spark-ignition, four-stroke model engines. The spark plug is fired on every stroke, resulting in a wasted spark in the cylinder when on its exhaust stroke.

Wasted Spark

The wasted spark system is an ignition system used in some four-stroke cycle internal combustion engines. In a wasted spark system, the spark plugs fire in pairs, with one plug in a cylinder on its compression stroke and the other plug in a cylinder on its exhaust stroke. The extra spark during the exhaust stroke has no effect and is thus "wasted". This design halves the number of components necessary in a typical ignition system, while the extra spark, against much reduced dielectric resistance, barely impacts the lifespan of modern ignition components. In a typical engine, it requires only about 2–3 kV to fire the cylinder on its exhaust stroke. The remaining coil energy is available to fire the spark plug in the cylinder on its compression stroke (typically about 8 to 12 kV).

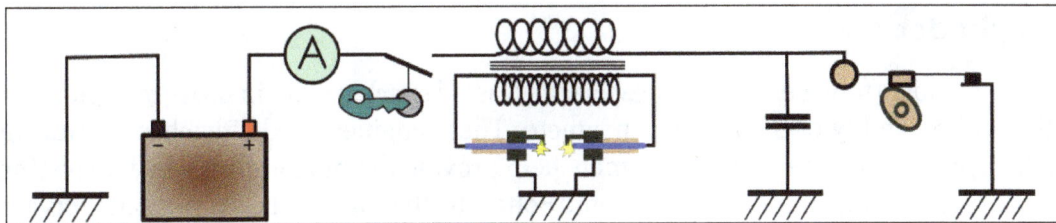
Ignition system of a flat-twin Citroën 2CV.

Advantages

Perhaps, the most significant advantage of the system, compared to a single coil and distributor systems, is that it eliminates the high-tension distributor. This significantly improves reliability, since many problems with a conventional system are caused by the distributor being affected by dampness from rain or condensation, dirt accumulation, and degradation of insulating materials with time. Although plug-top coil systems would later offer this same advantage, they were not available for another 30 years. Plug-top systems increase the number of coils required, increase the heat that these coils must survive and thus require more sophisticated and expensive materials to survive routine usage.

Timing Signal

Wasted spark systems still require a timing signal from the crankshaft. In a conventional four-stroke engine, this signal must also observe the phase of the camshaft relative to the crankshaft, so contact breakers are normally driven from the camshaft and distributor drive. With a wasted spark, the crankshaft can be used instead, as the system fires on both revolutions. It simplifies the mechanical arrangements since there is no distributor drive and the contact breaker cam can be fixed to the crankshaft. Although, a 2:1 reduction gear is still required to operate the camshaft and valves, the precision of this drive is now less critical (ignition timing is more critical for engine performance than valve timing): engines continue to run adequately even with worn camshaft drives and imprecise timing.

Practical Examples of 'Wasted Spark'

This system has been widely used, including such engines as the MG MG6 1.8T engine; Mitsubishi Evolution 4G63 engine, Mercedes-Benz inline sixes (M104.94x, M104.98x, M104.99x); Buick 3800 engines (LN3 and newer); Harley-Davidson V-Twin; air-cooled BMW Motorcycles; 1948 Citroën 2CV, Mazda B engines; Chrysler V10s; GY6 engine; Volkswagen Mark 3 2.8 VR6 (other than 2.0); Saturn Corporation 4 cylinders; Toyota 5VZ-FE V6s; Toyota 5E-FE and Chrysler 1.8, 2.0 & 2.4 engines. Some Ford engines also do. Many Honda and Kawasaki motorcycle and PWC engines also follow a similar design, to allow for a smaller number of more powerful coils to replace a larger number of smaller coils in the same limited space.

In practical use, a V-6 engine would only need three coil packs instead of six. Each individual coil fires the spark plugs in two cylinders simultaneously, the spark plug in one cylinder on a compression stroke where the power comes from, and the spark plug in the other cylinder on an exhaust stroke. Coils in a wasted spark system may be in pack form, or they may be in Coil-On-Plug (COP) form, with a spark plug cable attached to each COP unit, which connects to another spark plug.

Single Cylinder Use

Most single cylinder [four-stroke] engines use the wasted spark system in order to capitalise on the simplicity and reliability of the flywheel magneto. These engines need a flywheel to run smoothly, and the heavy current-generating magnets help provide the momentum while delivering a zero-maintenance drive to the ignition system. Bolted to the end of the crankshaft, this flywheel rotates twice for each compression stroke.

As well as the build-weight and maintenance advantages of this system, there is a tuning advantage. Mounted directly on the end of the crankshaft, all the stress that spark generation would otherwise place on the camshaft chain (or another special purpose half-speed drive) is avoided, while there is virtually none of the strain that necessarily degrades the ignition timing in systems relying on chains or gears.

Unlike the multi-cylinder systems noted above (which fire two plugs simultaneously from a double-ended coil) the coil in this system has only a single HT lead running to the single plug. The flywheel magneto provides other services in, for instance, small motorcycles, as it can easily be built to provide direct-current battery-charging power at almost no additional cost or weight.

Effect on Component Life

In modern conditions, this method has a very small impact on the length on the service intervals of the vehicle and the longevity of individual components. Modern ignition systems do not have breaker points, which have been almost entirely replaced by electronic systems. Modern ignition coils outlast most other components of the vehicle and modern spark plugs have excellent service life, though there is a slight-difference between the two plugs as to erosion suffered at the center electrode. Because the spark jumps in opposite directions on the companion plugs, one bank will erode the center electrode more, and the opposite bank will erode the ground electrode more. Spark plugs used in wasted spark systems should have precious metals, such as platinum and/or iridium, on both the central and ground electrodes in order to increase the average service interval time before replacement is needed.

Twin Plug Combustion

Since the earliest gasoline engines, some have used twin spark systems. These have two spark plugs in each cylinder. Each set of plugs is supplied separately. There may be several reasons for this: reliability (typically aircraft), starting, or better combustion performance by initiating the flamefront at opposing points simultaneously (e.g. Alfa Romeo). These are not considered as wasted spark systems, as their sparks all take place after the useful compression stroke, rather than "wasted" in the exhaust stroke.

References

- Carburetor, technology: britannica.com, Retrieved 19 April, 2019

- Hollembeak, barry (2005). Classroom manual for automotive fuels and emissions. Cengage learning. P. 154. Retrieved june 12, 2012

- Requirement-of-fuel-filter-in-engine: engihub.com, Retrieved 5 February, 2019

- "Misfire from driving through "wave of water" likely result of wet distributor cap — a problem that›s not costly to fix". Post and courier. Retrieved 2016-02-12

- Types-of-ignition-system: theengineerspost.com, Retrieved 26 July, 2019

5

Exhaust and Cooling Systems

The exhaust system in an automobile comprises of the piping which is used to guide reaction exhaust gases away from a controlled combustion inside an engine. The cooling system is employed to keep the temperature of the engine from exceeding the limits imposed by the needs of efficiency and safety. This chapter discusses in detail the diverse aspects of exhaust and cooling systems in an internal combustion engine.

Exhaust System of an Internal Combustion Engine

An exhaust system of the internal combustion engine comprises upstream main paths for cylinders that are attached to a side of a cylinder head and extend towards a side of the engine, and are connected to the respective cylinders; a downstream main path in which the upstream main paths join so as to become one flow path; a main catalytic converter provided on the downstream main path; bypasses that are split from the upstream main paths or the downstream main path; a bypass catalytic converter that is provided on the bypass; and flow path switching valves that open and close the upstream main paths so that exhaust discharged from the cylinders flows into the bypass. The bypass catalytic converter is provided below the upstream main paths.

In a conventional system, a main catalytic converter is arranged on the downstream side of an exhaust system, such as below a vehicle body floor. In such a system, a sufficient exhaust purification cannot be expected after a cold start of the internal combustion engine and until the temperature of the catalytic converter rises so that the converter is activated. In addition, the closer to the upstream side of the exhaust system the catalytic converter is, namely to the internal combustion engine side, the more problems there are with decreased durability due to the thermal deterioration of the catalyst of the converter.

An exhaust system has been proposed in which a bypass is provided in parallel to an upstream side portion of the main path having the main catalytic converter, and another bypass catalytic converter is provided on the bypass, and a switching valve for switching these paths are provided there between so that the exhaust is guided to the bypass immediately after a cold start. With this structure, the bypass catalytic converter is positioned on the upstream side of the main catalytic converter in the exhaust system and is activated at a relatively early stage so that exhaust purification can be started from the earlier stage.

According to the conventional exhaust system, the bypass splits from the main path, downstream of the confluence point of the exhaust manifold. In other words, the main path and the bypass are parallel, downstream of the confluence point at which the exhaust paths extending from respective cylinders of a multiple cylinder internal combustion engine are joined together, so that the device becomes large, and in particular, when the bypass catalytic converter is provided close to the internal combustion engine, it is difficult to provide the converter in the engine room of the vehicle.

Drawings

Other features and advantages of the present exhaust system will be apparent from the ensuing description, taken in conjunction with the accompanying drawings.

While the claims are not limited to the illustrated embodiments, an appreciation of various aspects of the exhaust system is best gained through a discussion of various examples thereof.

Description of the exhaust system which is applied to an inline 4-cylinder internal combustion engine will be given below as an example, by referring to drawings.

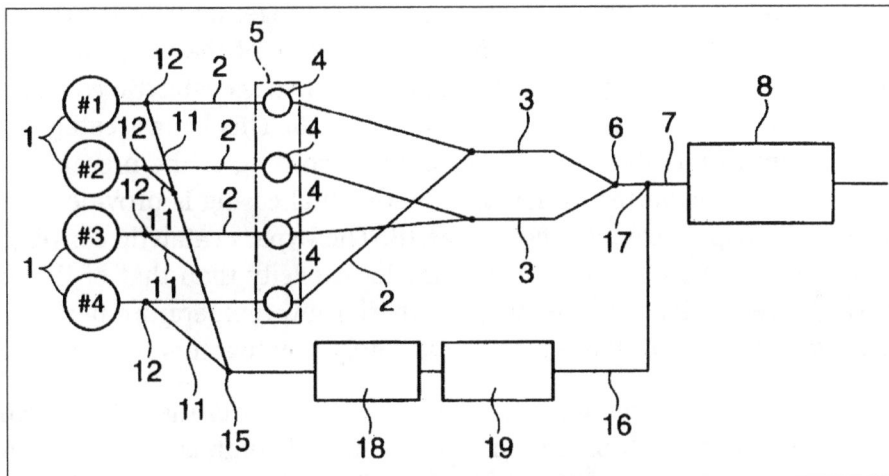

The structure of the entire exhaust system is described.

The cylinders 1 (1 to 4) that are arranged in a line are connected to respective upstream paths 2. Among the four cylinders, the upstream main path 2 for the cylinder 1 and the upstream main path 2 for the cylinder 4, in which the exhaust processes are not continued, are joined together so as to become a single middle main path 3, and similarly, the upstream main path 2 for the cylinder 2 and the upstream main path 2 for the cylinder 3, in which the exhaust processes are not continued, are joined together so as to become a single middle main path 3. Here, in each of the upstream main paths 2, a flow path switching valve 4 is provided. These flow path switching valves 4 are closed during a cold period, and further the four flow path switching valves 4 are provided as a single valve unit 5 so that all of the cylinders are opened and closed at the same time.

The two middle main paths 3 that are provided, downstream of the flow path switching valves 4, are joined together at a confluence point 6, so as to become a single downstream side main path 7. A main catalytic converter 8 is provided on the downstream main path 7. The main catalytic converter 8 has catalysts such as three-way catalyst and an HC trap catalyst. This main catalytic converter 8 has a large capacity and is arranged on undersurface of the vehicle floor. The upstream main paths 2, the middle main paths 3, the downstream main path 7, and the main catalytic converter 8 form a main path where the exhaust flows during the normal operation. These main paths have a pipe layout in which they are joined together in, as known as a "four-two-one form" in the inline 4-cylinder internal combustion engine, and therefore, the filling efficiency is improved by the dynamic exhaust effect.

On the other hand, an upstream bypass 11 is split from each of the upstream main paths 2 as a bypass. These upstream bypasses 11 have a sufficiently smaller cross-sectional path area than that of the upstream main path 2. A confluence point 12, which is located at the upstream end of each of the paths, is positioned as upstream as possible on the upstream main path 2. The upstream bypasses 11 for the four cylinders are eventually joined together so as to become a single downstream bypass 16 at a confluence point 15. It is important that the entire length of the bypass (the total sum of the bypasses for each cylinder) is short so that the thermal capacity of the pipe themselves and the heat loss area to the external atmosphere are small. As described later, the upstream bypasses 11 for the cylinders 2, 3, and 4 are connected at an approximately right angle to the upstream bypass 11 for the cylinder 1, which extends from the confluence point 12 of the cylinder 1 in the direction of the cylinder arrangement.

The downstream end of the downstream bypass 16 is joined together with the downstream main path 7 at a confluence point 17, which is on the upstream side of the main catalytic converter 8 provided on the downstream main path 7. Additionally, a bypass catalytic converter 18 using a three-way catalyst is provided on the downstream bypass 16. This bypass catalytic converter 18 is provided as upstream as possible on the bypass 16. According to the present embodiment, a secondary bypass catalytic converter 19 having an individual casing is provided in series on the downstream side of the bypass catalytic converter 18. The bypass catalytic converter 18 and the secondary bypass catalytic converter 19 have a smaller capacity than that of the main catalytic converter 8 in which preferably, a catalyst with a superior low temperature performance is used. Different catalysts may be used for these two bypass catalytic converters 18 and 19.

The figure is merely an explanatory diagram to illustrate the flow of the exhaust, which does not show the accurate position of each part in an actual internal combustion engine. Although, the bypass catalytic converter 18 is shown in parallel to the main converter 8, the bypass catalyst converter 18 is provided approximately at right angle with respect to the main converter 8, and is provided in the cylinder arrangement direction.

According to the exhaust system having the above-mentioned structure, when the engine temperature or the exhaust temperature is low after a cold start, the flow path switching valves 4 are closed by the an appropriate actuator, so that the main path is covered. Therefore, all the exhaust discharged from the cylinders 1 flows through the bypass catalytic converter 18 from the confluence points 12 and the upstream bypasses 11. The bypass catalytic converter 18 is positioned on the upstream side of the exhaust system, namely at a position close to the cylinders 1 so that it is compact, and it can be activated immediately and the exhaust purification is started at an early stage. In addition, at this time, the flow path switching paths 4 are closed so that the upstream main paths 2 for the respective cylinders 1 are disconnected from each other. Therefore, they prevent the exhaust discharged from the cylinders from flowing into the upstream main path 2 for other cylinders, and therefore the reduction of the exhaust temperature due to this phenomenon is certainly avoided. At a minimum, the number of the upstream portions of the bypasses is the same as that of the cylinders, and they are split on the upstream side of the confluence point of the upstream main path. Therefore, it is possible to position the bypass catalytic converter on the upstream side without restriction as to the position of the confluence point of the main path. In addition, since the splitting points thereof on the bypass side are close to the cylinders, the exhaust flows into the bypass without being relatively affected by the cooling effect due to the thermal capacity of the main path (exhaust manifold).

After the engine is warmed up, the engine temperature or the exhaust temperature become sufficiently high, and then the flow path switching valves are opened. The exhaust discharged from the cylinders mainly flows from the upstream main paths to the downstream main path and then flows through the main catalytic converter. Although at this time, the bypass is not particularly blocked, since the cross-sectional area of the bypass is smaller than the main path and the bypass catalytic converter and the secondary bypass catalytic converter are positioned in the middle, a majority of the exhaust flows through the main path and barely flows to the bypass due to the difference in the air flow resistance thereof, so that the thermal deterioration of the bypass catalytic converter is sufficiently restrained. In addition, the bypass is not completely blocked, so that during a high-speed high-load period when the amount of the exhaust is large, part of the exhaust flows through the bypass, thereby avoiding the reduction of the filling efficiency due to the back pressure.

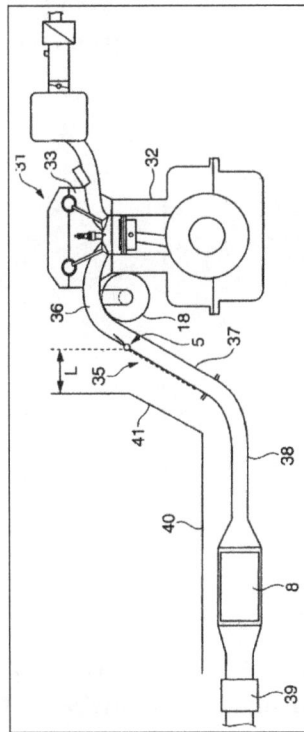

Figure above shows the detailed structure of the exhaust system which is installed in a vehicle. The inline 4-cylinder internal combustion engine 31 that comprises a cylinder block 32 and a cylinder head 33, is mounted in the engine room at the front portion of the vehicle in the so-called transverse manner, and an exhaust manifold 35 having four branch pipes 36, which are equivalent to the upstream main paths 2, is mounted on a side of the cylinder head 33 towards the rear side of the vehicle. The exhaust manifold 35 comprises a valve unit 5 in a middle portion thereof, in which the valve unit 5 has the flow path switching valves 4. The pipes are joined together so as to become one flow path as an outlet pipe 37. Additionally, a front tube 38 having the main catalytic converter

8, which is equivalent to the downstream main path 7, is connected to the outlet pipe 37. This exhaust system, as a whole, extends from the internal combustion engine 31 to the rear side of the vehicle. A silencer 39 is provided, downstream of the main catalytic converter 8.

Here, the main catalytic converter 8 is provided on the undersurface of the vehicle floor panel 40 with the silencer 39. In addition, the exhaust manifold 35 extends obliquely downward from the height of the cylinder head 33 to the height of the underfloor, along the dash panel 41 of the vehicle body. In particular, the upstream portion of each of the branch pipes 36, which are connected to the cylinder head 33, has an arched shape so that it smoothly heads downward. Additionally, a bypass catalytic converter 18 is provided in a space below the branch pipes 36 of the exhaust manifold 35 as high as possible between the exhaust manifold 35 and a side of the cylinder block 32. The bypass catalytic converter 18, which has an approximately cylindrical shape, has the inlet and outlet portions, at both ends thereof. The inlet portion is positioned below a branch pipe at one end of the internal combustion engine 31, and the outlet portion is positioned below a branch pipe at the other end of the internal combustion engine 31. The axis of the flow extends along the cylinder arrangement direction of the internal combustion engine 31 (in the direction of the crankshaft). Thus, the bypass catalytic converter 18 with the approximately cylinder shape is surrounded by the branch pipes 36 around the upper arch portion thereof. A space L is provided between the exhaust manifold 35 and the dash panel 41 in order to prevent thermal damage and to secure collision safety.

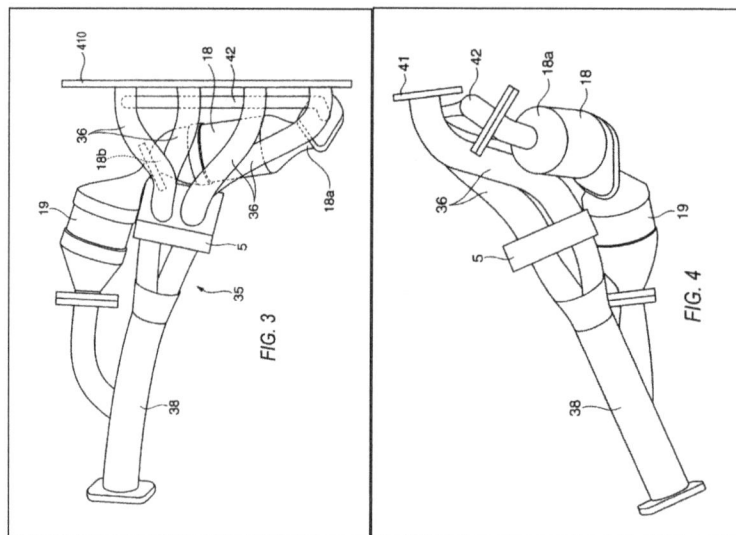

FIG. 3

FIG. 4

Figures above show the detailed structure of the above-mentioned exhaust manifold 35 in which first figure is a plan view and second figure is a side view thereof. The valve unit 5 has a flow path switching valve 4 around each of apexes of the square, and each of the four branch pipes 36 is connected to the flange 410 for attachment of the cylinder head at the upstream end thereof, and the downstream end thereof are connected to the valve unit 5. The approximately cylinder-shaped bypass catalytic converter 18 is provided below the four branch pipes 36. The bypass pipe 42 that is equivalent to the upstream bypass 11 extending from the cylinder 1, extends below the above-mentioned branch pipes 36 in parallel to the flange 41, that is, in the direction of the cylinder attachment. This bypass pipe 42 is, as shown in second figure, connected to the respective upstream ends of the branch pipes 36. The end of the bypass pipe 42 that extends from one end of the cylinder (for example the 1 cylinder) to the other end (for example the 4 cylinder) in its attachment direction is bent back in a U-turn shape and connected to the inlet portion 18 *a* of the bypass catalytic converter 18. As described above, the inlet portion 18 *a* of the bypass catalytic converter 18 that is arranged in the cylinder arrangement direction is positioned near the cylinder 4 and an outlet portion 18 *b* on the other end is positioned near the cylinder 1. In other words, the bypass catalytic converter 18 is positioned below the branch pipes 36 so that the space in the direction of the cylinder arrangement direction, in which the four branch pipes 36 are arranged, can be used as much as possible. The secondary bypass catalytic converter 19 is connected to the outlet portion 18 *b* in a bent shape towards the rear side of the vehicle. The secondary bypass catalytic converter 19 is provided on a side of the valve unit 5 and below the valve unit 5.

The bypass catalytic converter 18 is provided below the exhaust manifold 35 along the cylinder arrangement direction, as described above, so that the dead space formed between the exhaust manifold 35 and the cylinder block 32 can be efficiently utilized. The main paths 2 (branch pipes 36 and front tube 38) that extend from the cylinder head 33 to a portion under the floor cannot be extremely bent because the path resistance at the maximum output has to be taken into account. Therefore, since the main path 2 is formed so as to curve smoothly and obliquely downwards from the cylinder head 33, a relatively large space is easily formed between a side of the cylinder block 32 and the exhaust manifold 35. Consequently, by using this space for the bypass catalytic converter 18, the entire system can become compact. In particular, since the bypass catalytic converter 18 is placed along the direction of the cylinder arrangement, the bypass catalytic converter 18 can have a sufficiently large capacity in a limited space. As described above, although when the bypass catalytic converter 18 is placed along the cylinder arrangement direction, the exhaust flow greatly bends multiple times, this path resistance of the bypass side does not affect the maximum output of the engine. Further, since a period in which the bypass is used is short, it does not cause a substantial problem. According to the above-mentioned structure, the bypass catalytic converter 18 is provided very close to the exhaust ports, so that the exhaust that exits from the exhaust port can immediately flow into the bypass catalytic converter 18 via the bypass pipes 42. Therefore, the thermal capacity of the exhaust path to the bypass catalytic converter 18 and the heat loss to the outside are minimized and the exhaust purification by the bypass catalytic converter 18 can be started at an early stage.

Muffler

Muffler, also called silencer, is a device through which the exhaust gases from an internal-combustion engine are passed to attenuate (reduce) the airborne noise of the engine. To be efficient as a sound reducer, a muffler must decrease the velocity of the exhaust gases and either absorb sound waves or cancel them by interference with reflected waves coming from the same source.

A typical sound-absorbing material used in a muffler is a thick layer of fine fibres; the fibres are caused to vibrate by the sound waves, thus converting the sound energy to heat. Mufflers that attenuate sound waves by interference are known as reactive mufflers. These devices generally separate the waves into two components that follow different paths and then come together again out of phase (out of step), thus canceling each other out and reducing the sound.

One important chamber is known as the Helmholtz resonator. This chamber is of a dimension carefully tuned to reflect and cancel sound waves of specified frequencies. In addition, the tubes can be perforated with small holes that allow the reflection and interference of sound waves of other frequencies. The result is the attenuation of sound across a range of desired frequencies.

Mufflers of the straight-through type have a single tube with small holes connecting with annular chambers that are frequently stuffed with a sound-absorbing material.

Working of Muffler

To say that a muffler muffles explains how this automotive component does without actually telling you very much at all. It's more about how it muffles sound. The inside of your muffler isn't empty – it's actually filled with tubes, channels, and holes. They're arranged in such a way that sound is directed through the system, losing energy as it travels.

Of course, that's an over-simplification. There's actually a lot of technology embodied in the humble automobile muffler. The interior of the muffler is designed not to dampen sound, but to combine sound waves and make them cancel one another out. To achieve this, the tubes, holes and channels inside must be perfectly aligned or the sound waves will simply bounce past one another, which wouldn't reduce the engine noise at all.

There are four sections in your muffler. The inlet is the part that attaches to the rest of the exhaust system, and where exhaust gas and sound enters. The resonator chamber is where a cancellation sound wave is created. Then there's the second section, which is where you'll find two perforated tubes that further cancel sound. Finally, there's the outlet, which emits both the little bit of sound remaining, as well as exhaust gas.

Catalytic Converter

A catalytic converter is an exhaust emission control device that reduces toxic gases and pollutants in exhaust gas from an internal combustion engine into less-toxic pollutants by catalyzing a redox reaction (an oxidation and a reduction reaction). Catalytic converters are usually used with internal combustion engines fueled by either gasoline or diesel—including lean-burn engines as well as kerosene heaters and stoves.

The first widespread introduction of catalytic converters was in the United States automobile market. To comply with the U.S. Environmental Protection Agency's stricter regulation of exhaust emissions, most gasoline-powered vehicles starting with the 1975 model year must be equipped with catalytic converters. These "two-way" converters combine oxygen with carbon monoxide (CO) and unburned hydrocarbons (HC) to produce carbon dioxide (CO_2) and water (H_2O). In 1981, two-way catalytic converters were rendered obsolete by "three-way" converters that also reduce oxides of nitrogen (NO_x); however, two-way converters are still used for lean-burn engines. This is because three-way-converters require either rich or stoichiometric combustion to successfully reduce NO_x.

Although catalytic converters are most commonly applied to exhaust systems in automobiles, they are also used on electrical generators, forklifts, mining equipment, trucks, buses, locomotives, and motorcycles. They are also used on some wood stoves to control emissions. This is usually in response to government regulation, either through direct environmental regulation or through health and safety regulations.

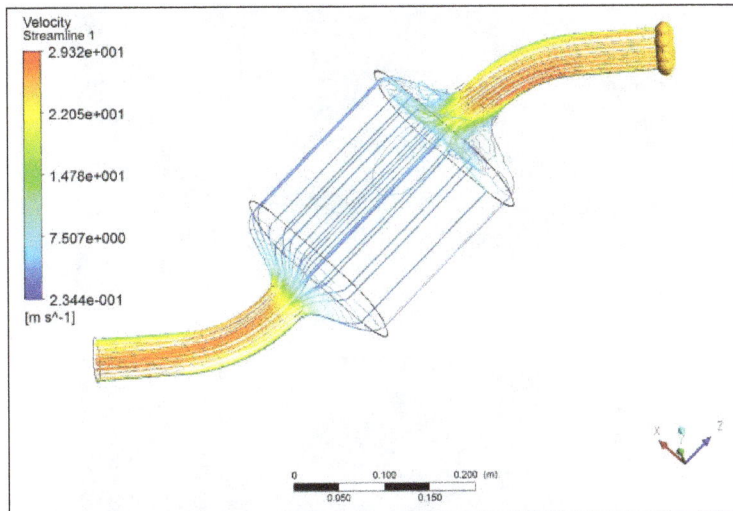

Simulation of flow inside a catalytic converter.

Construction

The catalytic converter's construction is as follows:

- The catalyst support or substrate: For automotive catalytic converters, the core is usually a ceramic monolith that has a honeycomb structure (commonly square, not hexagonal). (Prior to the mid 1980s, the catalyst material was deposited on a packed bed of pellets, especially in early GM applications.) Metallic foil monoliths made of Kanthal (FeCrAl) are used in applications where particularly high heat resistance is required. The substrate is structured to produce a large surface area. The cordierite ceramic substrate used in most catalytic converters was invented by Rodney Bagley, Irwin Lachman, and Ronald Lewis at Corning Glass, for which they were inducted into the National Inventors Hall of Fame in 2002.

- The washcoat: A washcoat is a carrier for the catalytic materials and is used to disperse the materials over a large surface area. Aluminum oxide, titanium dioxide, silicon dioxide, or a mixture of silica and alumina can be used. The catalytic materials are suspended in the washcoat prior to applying to the core. Washcoat materials are selected to form a rough, irregular surface, which greatly increases the surface area compared to the smooth surface of the bare substrate. This in turn maximizes the catalytically active surface available to react with the engine exhaust. The coat must retain its surface area and prevent sintering of the catalytic metal particles even at high temperatures (1000 °C).

- Ceria or ceria-zirconia: These oxides are mainly added as oxygen storage promoters.

- The catalyst itself is most often a mix of precious metal: Platinum is the most active catalyst and is widely used, but is not suitable for all applications because of unwanted additional reactions and high cost. Palladium and rhodium are two other precious metals used. Rhodium is used as a reduction catalyst, palladium is used as an oxidation catalyst, and platinum is used both for reduction and oxidation. Cerium, iron, manganese, and nickel are also used, although each has limitations. Nickel is not legal for use in the European Union because of its reaction with carbon monoxide into toxic nickel tetracarbonyl. Copper can be used everywhere except Japan.

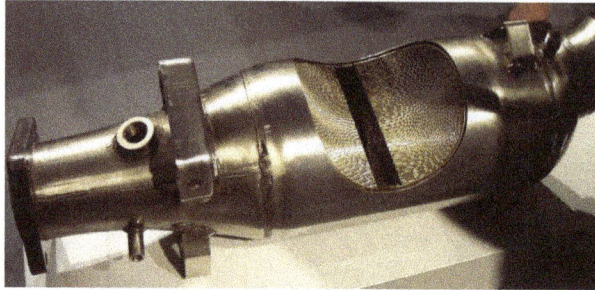
Cutaway of a metal-core converter.

Ceramic-core converter.

Upon failure, a catalytic converter can be recycled into scrap. The precious metals inside the converter, including platinum, palladium, and rhodium, are extracted.

Placement of Catalytic Converters

Catalytic converters require a temperature of 800 degrees Fahrenheit (426 °C) to efficiently convert harmful exhaust gases into inert gases, such as carbon dioxide and water vapor. Therefore, the first catalytic converters were placed close to the engine, to ensure fast heating. However, such placement can cause several problems. One of these is vapor lock.

As an alternative, catalytic converters were moved to a third of the way back from the engine, and were then placed underneath the vehicle.

Types

Two-way

A 2-way (or "oxidation", sometimes called an "oxi-cat") catalytic converter has two simultaneous tasks:

- Oxidation of carbon monoxide to carbon dioxide: $2\,CO + O_2 \rightarrow 2\,CO_2$.

- Oxidation of hydrocarbons (unburnt and partially burned fuel) to carbon dioxide and water: $C_xH_{2x+2} + [(3x+1)/2]\,O_2 \rightarrow x\,CO_2 + (x+1)\,H_2O$ (a combustion reaction).

This type of catalytic converter is widely used on diesel engines to reduce hydrocarbon and carbon monoxide emissions. They were also used on gasoline engines in American- and Canadian-market automobiles until 1981. Because of their inability to control oxides of nitrogen, they were superseded by three-way converters.

Three-way

Three-way catalytic converters (TWC) have the additional advantage of controlling the emission of nitric oxide (NO) and nitrogen dioxide (NO_2) (both together abbreviated with NO_x and nitrous oxide (N_2O)), which are precursors to acid rain and smog.

Since 1981, "three-way" (oxidation-reduction) catalytic converters have been used in vehicle emission control systems in the United States and Canada; many other countries have also adopted stringent vehicle emission regulations that in effect require three-way converters on gasoline-powered vehicles. The reduction and oxidation catalysts are typically contained in a common housing; however, in some instances, they may be housed separately. A three-way catalytic converter has three simultaneous tasks:

- Reduction of nitrogen oxides to nitrogen (N_2):
  - $2\,CO + 2\,NO \rightarrow 2\,CO_2 + N_2$
  - $hydrocarbon + NO \rightarrow CO_2 + H_2O + N_2$
  - $2\,H_2 + 2\,NO \rightarrow 2\,H_2O + N_2$
- Oxidation of carbon monoxide to carbon dioxide:
  - $2\,CO + O_2 \rightarrow 2\,CO_2$
- Oxidation of unburnt hydrocarbons (HC) to carbon dioxide and water, in addition to the above NO reaction:
  - $hydrocarbon + O_2 \rightarrow H_2O + CO_2$

These three reactions occur most efficiently when the catalytic converter receives exhaust from an engine running slightly above the stoichiometric point. For gasoline combustion, this ratio is between 14.6 and 14.8 parts air to one part fuel, by weight. The ratio for autogas (or liquefied petroleum gas LPG), natural gas, and ethanol fuels can be significantly different for each, notably so with oxygenated or alcohol based fuels, with e85 requiring approximately 34% more fuel to reach stoic, requiring modified fuel system tuning and components when using those fuels. In general, engines fitted with 3-way catalytic converters are equipped with a computerized closed-loop feedback fuel injection system using one or more oxygen sensors, though early in the deployment of three-way converters, carburetors equipped with feedback mixture control were used.

Three-way converters are effective when the engine is operated within a narrow band of air-fuel ratios near the stoichiometric point, such that the exhaust gas composition oscillates between rich (excess fuel) and lean (excess oxygen). Conversion efficiency falls very rapidly when the engine is operated outside of this band. Under lean engine operation, the exhaust contains excess oxygen, and the reduction of NO_x is not favored. Under rich conditions, the excess fuel consumes all of the available oxygen prior to the catalyst, leaving only oxygen stored in the catalyst available for the oxidation function.

Closed-loop engine control systems are necessary for effective operation of three-way catalytic converters because of the continuous balancing required for effective NO_x reduction and HC oxidation. The control system must prevent the NO_x reduction catalyst from becoming fully oxidized, yet replenish the oxygen storage material so that its function as an oxidation catalyst is maintained.

Three-way catalytic converters can store oxygen from the exhaust gas stream, usually when the air–fuel ratio goes lean. When sufficient oxygen is not available from the exhaust stream, the stored oxygen is released and consumed. A lack of sufficient oxygen occurs either when oxygen derived from NO_x reduction is unavailable or when certain maneuvers such as hard acceleration enrich the mixture beyond the ability of the converter to supply oxygen.

Unwanted Reactions

Unwanted reactions can occur in the three-way catalyst, such as the formation of odoriferous hydrogen sulfide and ammonia. Formation of each can be limited by modifications to the washcoat and precious metals used. It is difficult to eliminate these byproducts entirely. Sulfur-free or low-sulfur fuels eliminate or reduce hydrogen sulfide.

For example, when control of hydrogen-sulfide emissions is desired, nickel or manganese is added to the washcoat. Both substances act to block the absorption of sulfur by the washcoat. Hydrogen sulfide forms when the washcoat has absorbed sulfur during a low-temperature part of the operating cycle, which is then released during the high-temperature part of the cycle and the sulfur combines with HC.

Diesel Engines

For compression-ignition (i.e., diesel) engines, the most commonly used catalytic converter is the diesel oxidation catalyst (DOC). DOCs contain palladium, platinum, and aluminium oxide, all of which catalytically oxidize the hydrocarbons and carbon monoxide with oxygen to form carbon dioxide and water.

- $2\,CO + O_2 \rightarrow 2\,CO_2$

- $C_xH_{2x+2} + [(3x+1)/2]\,O_2 \rightarrow x\,CO_2 + (x+1)\,H_2O$

These converters often operate at 90 percent efficiency, virtually eliminating diesel odor and helping reduce visible particulates (soot). These catalysts do not reduce NO_x because any reductant present would react first with the high concentration of O_2 in diesel exhaust gas.

Reduction in NO_x emissions from compression-ignition engines has previously been addressed by the addition of exhaust gas to incoming air charge, known as exhaust gas recirculation (EGR).

In 2010, most light-duty diesel manufacturers in the U.S. added catalytic systems to their vehicles to meet new federal emissions requirements. There are two techniques that have been developed for the catalytic reduction of NO_x emissions under lean exhaust conditions: selective catalytic reduction (SCR) and the lean NO_x trap or NO_x adsorber.

Instead of precious metal-containing NO_x absorbers, most manufacturers selected base-metal SCR systems that use a reagent such as ammonia to reduce the NO_x into nitrogen. Ammonia is supplied to the catalyst system by the injection of urea into the exhaust, which then undergoes thermal decomposition and hydrolysis into ammonia. The urea solution is also referred to as Diesel Exhaust Fluid (DEF).

Diesel exhaust contains relatively high levels of particulate matter (PM). Catalytic converters do not remove PM so particulates are cleaned up by a soot trap or diesel particulate filter (DPF). In the U.S., all on-road light, medium and heavy-duty vehicles powered by diesel and built after January 1, 2007, must meet diesel particulate emission limits, meaning that they effectively have to be equipped with a 2-way catalytic converter and a diesel particulate filter. Note that this applies only to the diesel engine used in the vehicle. As long as the engine was manufactured before January 1, 2007, the vehicle is not required to have the DPF system. This led to an inventory runup by engine manufacturers in late 2006 so they could continue selling pre-DPF vehicles well into 2007.

Lean-burn Spark-ignition Engines

For lean-burn spark-ignition engines, an oxidation catalyst is used in the same manner as in a diesel engine. Emissions from lean burn spark ignition engines are very similar to emissions from a diesel compression ignition engine.

Installation

Many vehicles have a close-coupled catalytic converter located near the engine's exhaust manifold. The converter heats up quickly, due to its exposure to the very hot exhaust gases, enabling it to reduce undesirable emissions during the engine warm-up period. This is achieved by burning off the excess hydrocarbons which result from the extra-rich mixture required for a cold start.

When catalytic converters were first introduced, most vehicles used carburetors that provided a relatively rich air-fuel ratio. Oxygen (O_2) levels in the exhaust stream were therefore generally insufficient for the catalytic reaction to occur efficiently. Most designs of the time therefore included secondary air injection, which injected air into the exhaust stream. This increased the available oxygen, allowing the catalyst to function as intended.

Some three-way catalytic converter systems have air injection systems with the air injected between the first (NO_x reduction) and second (HC and CO oxidation) stages of the converter. As in two-way converters, this injected air provides oxygen for the oxidation reactions. An upstream air injection point, ahead of the catalytic converter, is also sometimes present to provide additional oxygen only during the engine warm up period. This causes unburned fuel to ignite in the exhaust tract, thereby preventing it reaching the catalytic converter at all. This technique reduces the engine runtime needed for the catalytic converter to reach its "light-off" or operating temperature.

Most newer vehicles have electronic fuel injection systems, and do not require air injection systems in their exhausts. Instead, they provide a precisely controlled air-fuel mixture that quickly and continually cycles between lean and rich combustion. Oxygen sensors monitor the exhaust oxygen content before and after the catalytic converter, and the engine control unit uses this information to adjust the fuel injection so as to prevent the first (NO_x reduction) catalyst from becoming oxygen-loaded, while simultaneously ensuring the second (HC and CO oxidation) catalyst is sufficiently oxygen-saturated.

Damage

Catalyst poisoning occurs when the catalytic converter is exposed to exhaust containing substances that coat the working surfaces, so that they cannot contact and react with the exhaust. The most notable contaminant is lead, so vehicles equipped with catalytic converters can run only on unleaded fuel. Other common catalyst poisons include sulfur, manganese (originating primarily from the gasoline additive MMT), and silicon, which can enter the exhaust stream if the engine has a leak that allows coolant into the combustion chamber. Phosphorus is another catalyst contaminant. Although phosphorus is no longer used in gasoline, it (and zinc, another low-level catalyst contaminant) was until recently widely used in engine oil antiwear additives such as zinc dithiophosphate (ZDDP). Beginning in 2004, a limit of phosphorus concentration in engine oils was adopted in the API SM and ILSAC GF-4 specifications.

Depending on the contaminant, catalyst poisoning can sometimes be reversed by running the engine under a very heavy load for an extended period of time. The increased exhaust temperature can sometimes vaporize or sublime the contaminant, removing it from the catalytic surface. However, removal of lead deposits in this manner is usually not possible because of lead's high boiling point.

Any condition that causes abnormally high levels of unburned hydrocarbons—raw or partially burnt fuel—to reach the converter will tend to significantly elevate its temperature, bringing the risk of a meltdown of the substrate and resultant catalytic deactivation and severe exhaust restriction. Usually the upstream components of the exhaust system (manifold/header assembly and associated clamps susceptible to rust/corrosion and fatigue e.g. the exhaust manifold splintering after repeated heat cycling), ignition system e.g. coil packs and primary ignition components (e.g. distributor cap, wires, ignition coil and spark plugs) and damaged fuel system components (fuel injectors, fuel pressure regulator, and associated sensors) - since 2006 ethanol has been used frequently with fuel blends where fuel system components which are not ethanol compatible can damage a catalytic converter - this also includes using a thicker oil viscosity not recommended by the manufacturer (especially with ZDDP content - this includes "high mileage" blends regardless if its conventional or synthetic oil), oil and coolant leaks (e.g. blown head gasket inclusive of engine overheating). Vehicles equipped with OBD-II diagnostic systems are designed to alert the driver to a misfire condition by means of illuminating the "check engine" light on the dashboard, or flashing it if the current misfire conditions are severe enough to potentially damage the catalytic converter.

Regulations

Emissions regulations vary considerably from jurisdiction to jurisdiction. Most automobile spark-ignition engines in North America have been fitted with catalytic converters since 1975, and the technology used in non-automotive applications is generally based on automotive technology.

Regulations for diesel engines are similarly varied, with some jurisdictions focusing on NO$_x$ (nitric oxide and nitrogen dioxide) emissions and others focusing on particulate (soot) emissions. This regulatory diversity is challenging for manufacturers of engines, as it may not be economical to design an engine to meet two sets of regulations.

Regulations of fuel quality vary across jurisdictions. In North America, Europe, Japan, and Hong Kong, gasoline and diesel fuel are highly regulated, and compressed natural gas and LPG (autogas) are being reviewed for regulation. In most of Asia and Africa, the regulations are often lax: in some places sulfur content of the fuel can reach 20,000 parts per million (2%). Any sulfur in the fuel can be oxidized to SO_2 (sulfur dioxide) or even SO_3 (sulfur trioxide) in the combustion chamber. If sulfur passes over a catalyst, it may be further oxidized in the catalyst, i.e., SO_2 may be further oxidized to SO_3. Sulfur oxides are precursors to sulfuric acid, a major component of acid rain. While it is possible to add substances such as vanadium to the catalyst washcoat to combat sulfur-oxide formation, such addition will reduce the effectiveness of the catalyst. The most effective solution is to further refine fuel at the refinery to produce ultra-low-sulfur diesel. Regulations in Japan, Europe, and North America tightly restrict the amount of sulfur permitted in motor fuels. However, the direct financial expense of producing such clean fuel may make it impractical for use in developing countries. As a result, cities in these countries with high levels of vehicular traffic suffer from acid rain, which damages stone and woodwork of buildings, poisons humans and other animals, and damages local ecosystems, at a very high financial cost.

Negative Aspects

Catalytic converters restrict the free flow of exhaust, which negatively affects vehicle performance and fuel economy, especially in older cars. Because early cars' carburetors were incapable of precise fuel-air mixture control, the cars' catalytic converters could overheat and ignite flammable materials under the car. A 2006 test on a 1999 Honda Civic showed that removing the stock catalytic converter netted a 3% increase in horsepower; a new metallic core converter only cost the car 1% horsepower, compared to no converter. To some performance enthusiasts, this modest increase in power for very little or no cost encourages the removal or "gutting" of the catalytic converter. In such cases, the converter may be replaced by a welded-in section of ordinary pipe or a flanged "test pipe", ostensibly meant to check if the converter is clogged, by comparing how the engine runs with and without the converter. This facilitates temporary reinstallation of the converter in order to pass an emission test. In many jurisdictions, it is illegal to remove or disable a catalytic converter for any reason other than its direct and immediate replacement. In the United States, for example, it is a violation of Section 203(a)(3)(A) of the 1990 amended Clean Air Act for a vehicle repair shop to remove a converter from a vehicle, or cause a converter to be removed from a vehicle, except in order to replace it with another converter, and Section 203(a)(3)(B) makes it illegal for any person to sell or to install any part that would bypass, defeat, or render inoperative any emission control system, device, or design element. Vehicles without functioning catalytic converters generally fail emission inspections. The automotive aftermarket supplies high-flow converters for vehicles with upgraded engines, or whose owners prefer an exhaust system with larger-than-stock capacity.

Warm-up Period

Vehicles fitted with catalytic converters emit most of their total pollution during the first five minutes of engine operation; for example, before the catalytic converter has warmed up sufficiently to be fully effective.

In 1995, Alpina introduced an electrically heated catalyst. Called "E-KAT," it was used in Alpina's B12 5,7 E-KAT based on the BMW 750i. Heating coils inside the catalytic converter assemblies are electrified just after the engine is started, bringing the catalyst up to operating temperature very quickly to qualify the vehicle for low emission vehicle (LEV) designation. BMW later introduced the same heated catalyst, developed jointly by Emitec, Alpina, and BMW, in its 750i in 1999.

Some vehicles contain a pre-cat, a small catalytic converter upstream of the main catalytic converter which heats up faster on vehicle start up, reducing the emissions associated with cold starts. A pre-cat is most commonly used by an auto manufacturer when trying to attain the Ultra Low Emissions Vehicle (ULEV) rating, such as on the Toyota MR2 Roadster.

Environmental Impact

Catalytic converters have proven to be reliable and effective in reducing noxious tailpipe emissions. However, they also have some shortcomings in use, and also adverse environmental impacts in production:

- An engine equipped with a three-way catalyst must run at the stoichiometric point, which means more fuel is consumed than in a lean-burn engine. This means approximately 10% more CO_2 emissions from the vehicle.

- Catalytic converter production requires palladium or platinum; part of the world supply of these precious metals is produced near Norilsk, Russia, where the industry (among others) has caused Norilsk to be added to *Time* magazine's list of most-polluted places.

- Pieces of catalytic converters, and the extreme heat of the converters themselves, can cause wildfires, especially in dry areas.

Theft

Because of the external location and the use of valuable precious metals including platinum, palladium, rhodium, and gold, catalytic converters are a target for thieves. The problem is especially common among late-model trucks and SUVs, because of their high ground clearance and easily removed bolt-on catalytic converters. Welded-on converters are also at risk of theft, as they can be easily cut off. The tools with which thieves quickly remove a catalytic converter, such as a portable reciprocating saw, can often damage other components of the car, such as the alternator, wiring or fuel lines, thus, there are dangerous consequences. Rising metal prices in the U.S. during the 2000s commodities boom led to a significant increase in converter theft, and unfortunately a catalytic converter can cost more than $1,000 to replace.

Diagnostics

Various jurisdictions now require on-board diagnostics to monitor the function and condition of the emissions-control system, including the catalytic converter. On-board diagnostic systems take several forms.

Temperature sensors are used for two purposes. The first is as a warning system, typically on two-way catalytic converters such as are still sometimes used on LPG forklifts. The function of the sensor is to warn of catalytic converter temperature above the safe limit of 750 °C (1,380 °F). More-recent catalytic-converter designs are not as susceptible to temperature damage and can withstand sustained temperatures of 900 °C (1,650 °F). Temperature sensors are also used to monitor catalyst functioning: usually two sensors will be fitted, with one before the catalyst and one after to monitor the temperature rise over the catalytic-converter core.

The oxygen sensor is the basis of the closed-loop control system on a spark-ignited rich-burn engine; however, it is also used for diagnostics. In vehicles with OBD II, a second oxygen sensor is fitted after the catalytic converter to monitor the O_2 levels. The O_2 levels are monitored to see the efficiency of the burn process. The on-board computer makes comparisons between the readings of the two sensors. The readings are taken by voltage measurements. If both sensors show the same output or the rear O_2 is "switching", the computer recognizes that the catalytic converter either is not functioning or has been removed, and will operate a malfunction indicator lamp and affect engine performance. Simple "oxygen sensor simulators" have been developed to circumvent this problem by simulating the change across the catalytic converter with plans and pre-assembled devices available on the Internet. Although, these are not legal for on-road use, they have been used with mixed results. Similar devices apply an offset to the sensor signals, allowing the engine to run a more fuel-economical lean burn that may, however, damage the engine or the catalytic converter.

NO_x sensors are extremely expensive and are in general used only when a compression-ignition engine is fitted with a selective catalytic-reduction (SCR) converter, or a NO_x absorber catalyst in a feedback system. When fitted to an SCR system, there may be one or two sensors. When one sensor is fitted it will be pre-catalyst; when two are fitted, the second one will be post-catalyst. They are used for the same reasons and in the same manner as an oxygen sensor; the only difference is the substance being monitored.

Internal Combustion Engine Cooling

Internal combustion engine cooling uses either air or liquid to remove the waste heat from an internal combustion engine. For small or special purpose engines, cooling using air from the atmosphere makes for a lightweight and relatively simple system. Watercraft can use water directly from the surrounding environment to cool their engines. For water-cooled engines on aircraft and surface vehicles, waste heat is transferred from a closed loop of water pumped through the engine to the surrounding atmosphere by a radiator.

Water has a higher heat capacity than air, and can thus move heat more quickly away from the engine, but a radiator and pumping system add weight, complexity, and cost. Higher-power engines generate more waste heat, but can move more weight, meaning they are generally water-cooled. Radial engines allow air to flow around each cylinder directly, giving them an advantage for air cooling over straight engines, flat engines, and V engines. Rotary engines have a similar configuration, but the cylinders also continually rotate, creating an air flow even when the vehicle is stationary.

Aircraft design more strongly favors lower weight and air-cooled designs. Rotary engines were popular on aircraft until the end of World War I, but had serious stability and efficiency problems. Radial engines were popular until the end of World War II, until gas turbine engines largely replaced them. Modern propeller-driven aircraft with internal-combustion engines are still largely air-cooled. Modern cars generally favor power over weight, and typically have water-cooled engines. Modern motorcycles are lighter than cars, and both cooling fluids are common. Some sport motorcycles were cooled with both air and oil (sprayed underneath the piston heads).

Heat engines generate mechanical power by extracting energy from heat flows, much as a water wheel extracts mechanical power from a flow of mass falling through a distance. Engines are inefficient, so more heat energy enters the engine than comes out as mechanical power; the difference is waste heat which must be removed. Internal combustion engines remove waste heat through cool intake air, hot exhaust gases, and explicit engine cooling.

Engines with higher efficiency have more energy leave as mechanical motion and less as waste heat. Some waste heat is essential: it guides heat through the engine, much as a water wheel works only if there is some exit velocity (energy) in the waste water to carry it away and make room for more water. Thus, all heat engines need cooling to operate.

Cooling is also needed because high temperatures damage engine materials and lubricants and becomes even more important in hot climates. Internal-combustion engines burn fuel hotter than the melting temperature of engine materials, and hot enough to set fire to lubricants. Engine cooling removes energy fast enough to keep temperatures low so the engine can survive.

Some high-efficiency engines run without explicit cooling and with only incidental heat loss, a design called adiabatic. Such engines can achieve high efficiency but compromise power output, duty cycle, engine weight, durability, and emissions.

Basic Principles

Most internal combustion engines are fluid cooled using either air (a gaseous fluid) or a liquid coolant run through a heat exchanger (radiator) cooled by air. Marine engines and some stationary engines have ready access to a large volume of water at a suitable temperature. The water may be used directly to cool the engine, but often has sediment, which can clog coolant passages, or chemicals, such as salt, that can chemically damage the engine. Thus, engine coolant may be run through a heat exchanger that is cooled by the body of water.

Most liquid-cooled engines use a mixture of water and chemicals such as antifreeze and rust inhibitors. The industry term for the antifreeze mixture is *engine coolant*. Some antifreezes use no water at all, instead using a liquid with different properties, such as propylene glycol or a combination of propylene glycol and ethylene glycol. Most "air-cooled" engines use some liquid oil cooling, to maintain acceptable temperatures for both critical engine parts and the oil itself. Most "liquid-cooled" engines use some air cooling, with the intake stroke of air cooling the combustion chamber. An exception is Wankel engines, where some parts of the combustion chamber are never cooled by intake, requiring extra effort for successful operation.

There are many demands on a cooling system. One key requirement is to adequately serve the entire engine, as the whole engine fails if just one part overheats. Therefore, it is vital that the cooling system keep *all* parts at suitably low temperatures. Liquid-cooled engines are able to vary the size of their passageways through the engine block so that coolant flow may be tailored to the needs of each area. Locations with either high peak temperatures (narrow islands around the combustion chamber) or high heat flow (around exhaust ports) may require generous cooling. This reduces the occurrence of hot spots, which are more difficult to avoid with air cooling. Air-cooled engines may also vary their cooling capacity by using more closely spaced cooling fins in that area, but this can make their manufacture difficult and expensive.

Only the fixed parts of the engine, such as the block and head, are cooled directly by the main coolant system. Moving parts such as the pistons, and to a lesser extent the crank and rods, must rely on the lubrication oil as a coolant, or to a very limited amount of conduction into the block and thence the main coolant. High performance engines frequently have additional oil, beyond the amount needed for lubrication, sprayed upwards onto the bottom of the piston just for extra cooling. Air-cooled motorcycles often rely heavily on oil-cooling in addition to air-cooling of the cylinder barrels.

Liquid-cooled engines usually have a circulation pump. The first engines relied on thermo-syphon cooling alone, where hot coolant left the top of the engine block and passed to the radiator, where it was cooled before returning to the bottom of the engine. Circulation was powered by convection alone. Other demands include cost, weight, reliability, and durability of the cooling system itself.

Conductive heat transfer is proportional to the temperature difference between materials. If engine metal is at 250 °C and the air is at 20 °C, then there is a 230 °C temperature difference for cooling. An air-cooled engine uses all of this difference. In contrast, a liquid-cooled engine might dump heat from the engine to a liquid, heating the liquid to 135 °C (Water's standard boiling point of 100 °C can be exceeded as the cooling system is both pressurised, and uses a mixture with antifreeze) which is then cooled with 20 °C air. In each step, the liquid-cooled engine has half the temperature difference and so at first appears to need twice the cooling area.

However, properties of the coolant (water, oil, or air) also affect cooling. As example, comparing water and oil as coolants, one gram of oil can absorb about 55% of the heat for the same rise in temperature (called the specific heat capacity). Oil has about 90% the density of water, so a given volume of oil can absorb only about 50% of the energy of the same volume of water. The thermal conductivity of water is about four times that of oil, which can aid heat transfer. The viscosity of oil can be ten times greater than water, increasing the energy required to pump oil for cooling, and reducing the net power output of the engine.

Comparing air and water, air has vastly lower heat capacity per gram and per volume (4000) and less than a tenth the conductivity, but also much lower viscosity (about 200 times lower: 17.4×10^{-6} Pa·s for air vs 8.94×10^{-4} Pa·s for water). Continuing the calculation from two paragraphs above, air cooling needs ten times of the surface area, therefore the fins, and air needs 2000 times the flow velocity and thus a recirculating air fan needs ten times the power of a recirculating water pump. Moving heat from the cylinder to a large surface area for air cooling can present problems such as difficulties manufacturing the shapes needed for good heat transfer and the space needed for free flow of a large volume of air. Water boils at about the same temperature desired for engine cooling. This has the advantage that it absorbs a great deal of energy with very little rise in temperature (called heat of vaporization), which is good for keeping things cool, especially for passing one stream of coolant over several hot objects and achieving uniform temperature. In contrast, passing air over several hot objects in series warms the air at each step, so the first may be over-cooled and the last under-cooled. However, once water boils, it is an insulator, leading to a sudden loss of cooling where steam bubbles form. Steam may return to water as it mixes with other coolant, so an engine temperature gauge can indicate an acceptable temperature even though local temperatures are high enough that damage is being done.

An engine needs different temperatures. The inlet including the compressor of a turbo and in the inlet trumpets and the inlet valves need to be as cold as possible. A countercurrent heat exchange with forced cooling air does the job. The cylinder-walls should not heat up the air before compression, but also not cool down the gas at the combustion. A compromise is a wall temperature of 90 °C. The viscosity of the oil is optimized for just this temperature. Any cooling of the exhaust and the turbine of the turbocharger reduces the amount of power available to the turbine, so the exhaust system is often insulated between engine and turbocharger to keep the exhaust gases as hot as possible.

The temperature of the cooling air may range from well below freezing to 50 °C. Further, while engines in long-haul boat or rail service may operate at a steady load, road vehicles often see widely varying and quickly varying load. Thus, the cooling system is designed to vary cooling so the engine is neither too hot nor too cold. Cooling system regulation includes adjustable baffles in the air flow (sometimes called 'shutters' and commonly run by a pneumatic 'shutterstat); a fan which operates either independently of the engine, such as an electric fan, or which has an adjustable clutch; a thermostatic valve or just 'thermostat' that can block the coolant flow when too cool. In addition, the motor, coolant, and heat exchanger have some heat capacity which smooths out temperature increase in short sprints. Some engine controls shut down an engine or limit it to half throttle if it overheats. Modern electronic engine controls adjust cooling based on throttle to anticipate a temperature rise, and limit engine power output to compensate for finite cooling.

Finally, other concerns may dominate cooling system design. As example, air is a relatively poor coolant, but air cooling systems are simple, and failure rates typically rise as the square of the number of failure points. Also, cooling capacity is reduced only slightly by small air coolant leaks. Where reliability is of utmost importance, as in aircraft, it may be a good trade-off to give up efficiency, longevity (interval between engine rebuilds), and quietness in order to achieve slightly higher reliability; the consequences of a broken airplane engine are so severe, even a slight increase in reliability is worth giving up other good properties to achieve it.

Air-cooled and liquid-cooled engines are both used commonly. Each principle has advantages and disadvantages, and particular applications may favor one over the other. For example, most cars and trucks use liquid-cooled engines, while many small airplane and low-cost engines are air-cooled.

Generalization Difficulties

It is difficult to make generalizations about air-cooled and liquid-cooled engines. Air-cooled diesel engines are chosen for reliability even in extreme heat, because air-cooling would be simpler and more effective at coping with the extremes of temperatures during the depths of winter and height of summer, than water cooling systems, and are often used in situations where the engine runs unattended for months at a time.

Similarly, it is usually desirable to minimize the number of heat transfer stages in order to maximize the temperature difference at each stage. However, Detroit Diesel two-stroke cycle engines commonly use oil cooled by water, with the water in turn cooled by air.

The coolant used in many liquid-cooled engines must be renewed periodically, and can freeze at ordinary temperatures thus causing permanent engine damage when it expands. Air-cooled engines do not require coolant service, and do not suffer damage from freezing, two commonly cited advantages for air-cooled engines. However, coolant based on propylene glycol is liquid to −55 °C, colder than is encountered by many engines; shrinks slightly when it crystallizes, thus avoiding damage; and has a service life over 10,000 hours, essentially the lifetime of many engines.

It is usually more difficult to achieve either low emissions or low noise from an air-cooled engine, two more reasons most road vehicles use liquid-cooled engines. It is also often difficult to build large air-cooled engines, so nearly all air-cooled engines are under 500 kW (670 hp), whereas large liquid-cooled engines exceed 80 MW (107000 hp) (Wärtsilä-Sulzer RTA96-C 14-cylinder diesel).

Air-cooling

A cylinder from an air-cooled aviation engine, a Continental C85. Notice the rows of fins on both the steel cylinder barrel and the aluminum cylinder head. The fins provide additional surface area for air to pass over the cylinder and absorb heat.

Cars and trucks using direct air cooling (without an intermediate liquid) were built over a long period from the very beginning and ending with a small and generally unrecognized technical change. Before World War II, water-cooled cars and trucks routinely overheated while climbing mountain roads, creating geysers of boiling cooling water. This was considered normal, and at the time, most noted mountain roads had auto repair shops to minister to overheating engines.

ACS (Auto Club Suisse) maintains historical monuments to that era on the Susten Pass where two radiator refill stations remain. These have instructions on a cast metal plaque and a spherical bottom watering can hanging next to a water spigot. The spherical bottom was intended to keep it from being set down and, therefore, be useless around the house, in spite of which it was stolen, as the picture shows.

During that period, European firms such as Magirus-Deutz built air-cooled diesel trucks, Porsche built air-cooled farm tractors, and Volkswagen became famous with air-cooled passenger cars. In the United States, Franklin built air-cooled engines.

For many years air cooling was favored for military applications as liquid cooling systems are more vulnerable to damage by shrapnel.

The Czech Republic–based company Tatra is known for their large displacement air-cooled V8 car engines; Tatra engineer Julius Mackerle published a book on it. Air-cooled engines are better adapted to extremely cold and hot environmental weather temperatures: you can see air-cooled engines starting and running in freezing conditions that seized water-cooled engines and continue working when water-cooled ones start producing steam jets. Air-cooled engines have may be an advantage from a thermodynamic point of view due to higher operating temperature. The worst problem met in air-cooled aircraft engines was the so-called "Shock cooling", when the airplane entered in a dive after climbing or level flight with throttle open, with the engine under no load while the airplane dives generating less heat, and the flow of air that cools the engine is increased, a catastrophic engine failure may result as different parts of engine have different temperatures, and thus different thermal expansions. In such conditions, the engine may seize, and any sudden change or imbalance in the relation between heat produced by the engine and heat dissipated by cooling may result in an increased wear of engine, as a consequence also of thermal expansion differences between parts of engine, liquid-cooled engines having more stable and uniform working temperatures.

Liquid Cooling

Today, most automotive and larger IC engines are liquid-cooled.

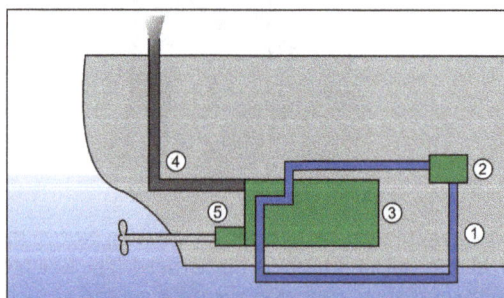

A fully closed IC engine cooling system.

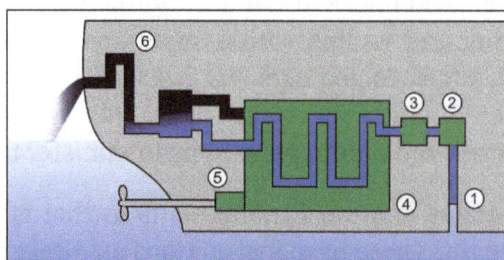

Open IC engine cooling system.

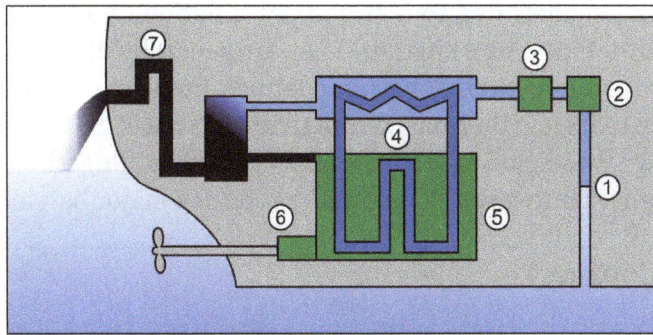
Semi-closed IC engine cooling system.

Liquid cooling is also employed in maritime vehicles (vessels, ...). For vessels, the seawater itself is mostly used for cooling. In some cases, chemical coolants are also employed (in closed systems) or they are mixed with seawater cooling.

Transition from Air Cooling

The change of air cooling to liquid cooling occurred at the start of World War II when the US military needed reliable vehicles. The subject of boiling engines was addressed, researched, and a solution found. Previous radiators and engine blocks were properly designed and survived durability tests, but used water pumps with a leaky graphite-lubricated "rope" seal (gland) on the pump shaft. The seal was inherited from steam engines, where water loss is accepted, since steam engines already expend large volumes of water. Because the pump seal leaked mainly when the pump was running and the engine was hot, the water loss evaporated inconspicuously, leaving at best a small rusty trace when the engine stopped and cooled, thereby not revealing significant water loss. Automobile radiators (or heat exchangers) have an outlet that feeds cooled water to the engine and the engine has an outlet that feeds heated water to the top of the radiator. Water circulation is aided by a rotary pump that has only a slight effect, having to work over such a wide range of speeds that its impeller has only a minimal effect as a pump. While running, the leaking pump seal drained cooling water to a level where the pump could no longer return water to the top of the radiator, so water circulation ceased and water in the engine boiled. However, since water loss led to overheat and further water loss from boil-over, the original water loss was hidden.

After isolating the pump problem, cars and trucks built for the war effort (no civilian cars were built during that time) were equipped with carbon-seal water pumps that did not leak and caused no more geysers. Meanwhile, air cooling advanced in memory of boiling engines... even though boil-over was no longer a common problem. Air-cooled engines became popular throughout Europe. After the war, Volkswagen advertised in the USA as not boiling over, even though new water-cooled cars no longer boiled over, but these cars sold well. But as air quality awareness rose in the 1960s, and laws governing exhaust emissions were passed, unleaded gas replaced leaded gas and leaner fuel mixtures became the norm. Subaru chose liquid-cooling for their EA series (flat) engine when it was introduced in 1966.

Low Heat Rejection Engines

A special class of experimental prototype internal combustion piston engines have been developed over several decades with the goal of improving efficiency by reducing heat loss. These engines are variously called adiabatic engines, due to better approximation of adiabatic expansion, low heat rejection engines, or high-temperature engines. They are generally diesel engines with combustion chamber parts lined with ceramic thermal barrier coatings. Some make use of titanium pistons and other titanium parts due to its low thermal conductivity and mass. Some designs are able to eliminate the use of a cooling system and associated parasitic losses altogether. Developing lubricants able to withstand the higher temperatures involved has been a major barrier to commercialization.

Radiator (Engine Cooling)

Radiators are heat exchangers used for cooling internal combustion engines, mainly in automobiles but also in piston-engined aircraft, railway locomotives, motorcycles, stationary generating plant or any similar use of such an engine.

Internal combustion engines are often cooled by circulating a liquid called *engine coolant* through the engine block, where it is heated, then through a radiator where it loses heat to the atmosphere, and then returned to the engine. Engine coolant is usually water-based, but may also be oil. It is common to employ a water pump to force the engine coolant to circulate, and also for an axial fan to force air through the radiator.

A typical engine coolant radiator used in an automobile.

Automobiles and Motorcycles

Coolant being poured into the radiator of an automobile.

In automobiles and motorcycles with a liquid-cooled internal combustion engine, a radiator is connected to channels running through the engine and cylinder head, through which a liquid (coolant) is pumped. This liquid may be water (in climates where water is unlikely to freeze), but is more commonly a mixture of water and antifreeze in proportions appropriate to the climate. Antifreeze itself is usually ethylene glycol or propylene glycol (with a small amount of corrosion inhibitor).

A typical automotive cooling system comprises:

- A series of galleries cast into the engine block and cylinder head, surrounding the combustion chambers with circulating liquid to carry away heat;

- A radiator, consisting of many small tubes equipped with a honeycomb of fins to dissipate heat rapidly, that receives and cools hot liquid from the engine;

- A water pump, usually of the centrifugal type, to circulate the coolant through the system;

- A thermostat to control temperature by varying the amount of coolant going to the radiator;

- A fan to draw cool air through the radiator.

The radiator transfers the heat from the fluid inside to the air outside, thereby cooling the fluid, which in turn cools the engine. Radiators are also often used to cool automatic transmission fluids, air conditioner refrigerant, intake air, and sometimes to cool motor oil or power steering fluid. Radiators are typically mounted in a position where they receive airflow from the forward movement of the vehicle, such as behind a front grill. Where engines are mid- or rear-mounted, it is common to mount the radiator behind a front grill to achieve sufficient airflow, even though this requires long coolant pipes. Alternatively, the radiator may draw air from the flow over the top of the vehicle or from a side-mounted grill. For long vehicles, such as buses, side airflow is most common for engine and transmission cooling and top airflow most common for air conditioner cooling.

Radiator Construction

Automobile radiators are constructed of a pair of metal or plastic header tanks, linked by a core with many narrow passageways, giving a high surface area relative to volume. This core is usually made of stacked layers of metal sheet, pressed to form channels and soldered or brazed together. For many years radiators were made from brass or copper cores soldered to brass headers. Modern radiators have aluminum cores, and often save money and weight by using plastic headers with gaskets. This construction is more prone to failure and less easily repaired than traditional materials.

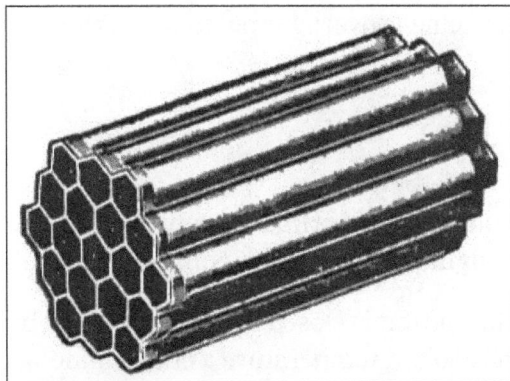

Honeycomb radiator tubes.

An earlier construction method was the honeycomb radiator. Round tubes were swaged into hexagons at their ends, then stacked together and soldered. As they only touched at their ends, this formed what became in effect a solid water tank with many air tubes through it.

Some vintage cars use radiator cores made from coiled tube, a less efficient but simpler construction.

Coolant Pump

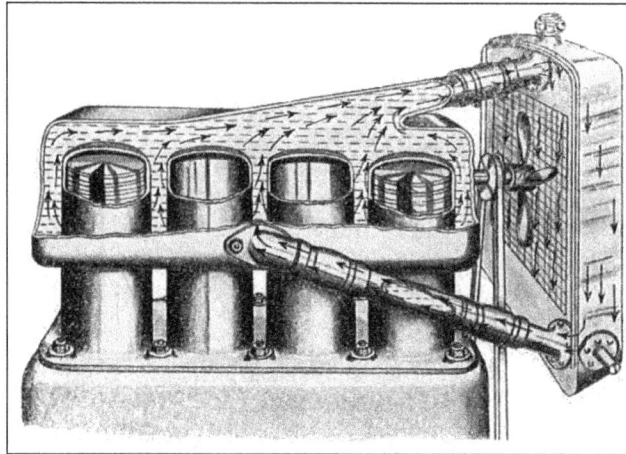

Thermosyphon cooling system of 1937, without circulating pump.

Radiators first used downward vertical flow, driven solely by a thermosyphon effect. Coolant is heated in the engine, becomes less dense, and so rises. As the radiator cools the fluid, the coolant becomes denser and falls. This effect is sufficient for low-power stationary engines, but inadequate for all but the earliest automobiles. All automobiles for many years have used centrifugal pumps to circulate the engine coolant because natural circulation has very low flow rates.

Heater

A system of valves or baffles, or both, is usually incorporated to simultaneously operate a small radiator inside the vehicle. This small radiator, and the associated blower fan, is called the heater core, and serves to warm the cabin interior. Like the radiator, the heater core acts by removing heat from the engine. For this reason, automotive technicians often advise operators to turn *on* the heater and set it to high if the engine is overheating, to assist the main radiator.

Temperature Control

Waterflow Control

The engine temperature on modern cars is primarily controlled by a wax-pellet type of thermostat, a valve which opens once the engine has reached its optimum operating temperature.

When the engine is cold, the thermostat is closed except for a small bypass flow so that the thermostat experiences changes to the coolant temperature as the engine warms up. Engine coolant is directed by the thermostat to the inlet of the circulating pump and is returned directly to the engine, bypassing the radiator. Directing water to circulate only through the engine allows the engine to reach optimum operating temperature as quickly as possible whilst avoiding localised "hot spots."

Once the coolant reaches the thermostat's activation temperature, it opens, allowing water to flow through the radiator to prevent the temperature rising higher.

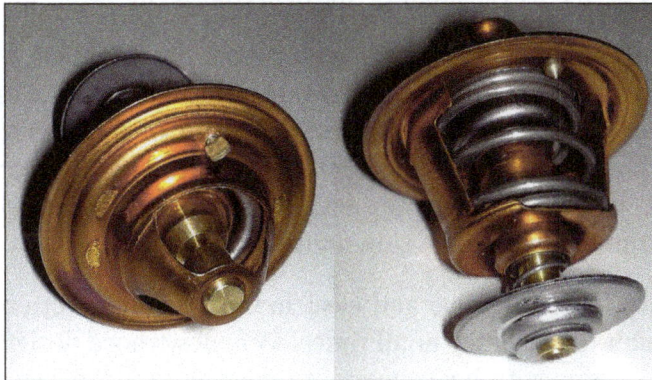

Car engine thermostat.

Once at optimum temperature, the thermostat controls the flow of engine coolant to the radiator so that the engine continues to operate at optimum temperature. Under peak load conditions, such as driving slowly up a steep hill whilst heavily laden on a hot day, the thermostat will be approaching fully open because the engine will be producing near to maximum power while the velocity of air flow across the radiator is low. (The velocity of air flow across the radiator has a major effect on its ability to dissipate heat.) Conversely, when cruising fast downhill on a motorway on a cold night on a light throttle, the thermostat will be nearly closed because the engine is producing little power, and the radiator is able to dissipate much more heat than the engine is producing. Allowing too much flow of coolant to the radiator would result in the engine being over cooled and operating at lower than optimum temperature, resulting in decreased fuel efficiency and increased exhaust emissions. Furthermore, engine durability, reliability, and longevity are sometimes compromised, if any components (such as the crankshaft bearings) are engineered to take thermal expansion into account to fit together with the correct clearances. Another side effect of over-cooling is reduced performance of the cabin heater, though in typical cases it still blows air at a considerably higher temperature than ambient.

The thermostat is therefore constantly moving throughout its range, responding to changes in vehicle operating load, speed and external temperature, to keep the engine at its optimum operating temperature.

On vintage cars you may find a bellows type thermostat, which has a corrugated bellows containing a volatile liquid such as alcohol or acetone. These types of thermostats do not work well at cooling system pressures above about 7 psi. Modern motor vehicles typically run at around 15 psi, which precludes the use of the bellows type thermostat. On direct air-cooled engines this is not a concern for the bellows thermostat that controls a flap valve in the air passages.

Airflow Control

Other factors influence the temperature of the engine, including radiator size and the type of radiator fan. The size of the radiator (and thus its cooling capacity) is chosen such that it can keep the engine at the design temperature under the most extreme conditions a vehicle is likely to encounter (such as climbing a mountain whilst fully loaded on a hot day).

Airflow speed through a radiator is a major influence on the heat it dissipates. Vehicle speed affects this, in rough proportion to the engine effort, thus giving crude self-regulatory feedback. Where an additional cooling fan is driven by the engine, this also tracks engine speed similarly.

Engine-driven fans are often regulated by a fan clutch from the drivebelt, which slips and reduces the fan speed at low temperatures. This improves fuel efficiency by not wasting power on driving the fan unnecessarily. On modern vehicles, further regulation of cooling rate is provided by either variable speed or cycling radiator fans. Electric fans are controlled by a thermostatic switch or the engine control unit. Electric fans also have the advantage of giving good airflow and cooling at low engine revs or when stationary, such as in slow-moving traffic.

Before the development of viscous-drive and electric fans, engines were fitted with simple fixed fans that drew air through the radiator at all times. Vehicles whose design required the installation of a large radiator to cope with heavy work at high temperatures, such as commercial vehicles and tractors would often run cool in cold weather under light loads, even with the presence of a thermostat, as the large radiator and fixed fan caused a rapid and significant drop in coolant temperature as soon as the thermostat opened. This problem can be solved by fitting a radiator blind (or radiator shroud) to the radiator that can be adjusted to partially or fully block the airflow through the radiator. At its simplest the blind is a roll of material such as canvas or rubber that is unfurled along the length of the radiator to cover the desired portion. Some older vehicles, like the World War I-era S.E.5 and SPAD S.XIII single-engined fighters, have a series of shutters that can be adjusted from the driver's or pilot's seat to provide a degree of control. Some modern cars have a series of shutters that are automatically opened and closed by the engine control unit to provide a balance of cooling and aerodynamics as needed.

These AEC Regent III RT buses are fitted with radiator blinds, seen here covering the lower half of the radiators.

Coolant Pressure

Because the thermal efficiency of internal combustion engines increases with internal temperature, the coolant is kept at higher-than-atmospheric pressure to increase its boiling point. A calibrated pressure-relief valve is usually incorporated in the radiator's fill cap. This pressure varies between models, but typically ranges from 4 to 30 psi (30 to 200 kPa).

As the coolant expands with increasing temperature, its pressure in the closed system must increase. Ultimately, the pressure relief valve opens, and excess fluid is dumped into an overflow container. Fluid overflow ceases when the thermostat modulates the rate of cooling to keep the temperature of the coolant at optimum. When the engine coolant cools and contracts (as conditions change or when the engine is switched off), the fluid is returned to the radiator through additional valving in the cap.

Cooling fan of radiator for prime mover of a VIA Rail locomotive.

Engine Coolant

Before World War II, engine coolant was usually plain water. Antifreeze was used solely to control freezing, and this was often only done in cold weather.

Development in high-performance aircraft engines required improved coolants with higher boiling points, leading to the adoption of glycol or water-glycol mixtures. These led to the adoption of glycols for their antifreeze properties.

Since, the development of aluminium or mixed-metal engines, corrosion inhibition has become even more important than antifreeze, and in all regions and seasons.

Boiling or Overheating

An overflow tank that runs dry may result in the coolant vaporizing, which can cause localized or general overheating of the engine. Severe damage can result, such as blown headgaskets, and warped or cracked cylinder heads or cylinder blocks. Sometimes there will be no warning, because the temperature sensor that provides data for the temperature gauge (either mechanical or electric) is exposed to air, not to the excessively hot coolant, providing a harmfully false reading.

Opening a hot radiator drops the system pressure, which may cause it to boil and eject dangerously hot liquid and steam. Therefore, radiator caps often contain a mechanism that attempts to relieve the internal pressure before the cap can be fully opened.

The invention of the automobile water radiator is attributed to Karl Benz. Wilhelm Maybach designed the first honeycomb radiator for the Mercedes 35hp.

Supplementary Radiators

It is sometimes necessary for a car to be equipped with a second, or auxiliary, radiator to increase the cooling capacity, when the size of the original radiator cannot be increased. The second radiator is plumbed in series with the main radiator in the circuit. This was the case when the Audi 100 was first turbocharged creating the 200.

Some engines have an oil cooler, a separate small radiator to cool the engine oil. Cars with an automatic transmission often have extra connections to the radiator, allowing the transmission fluid to transfer its heat to the coolant in the radiator. These may be either oil-air radiators, as for a smaller version of the main radiator. More simply they may be oil-water coolers, where an oil pipe is inserted inside the water radiator. Though the water is hotter than the ambient air, its higher thermal conductivity offers comparable cooling (within limits) from a less complex and thus cheaper and more reliable oil cooler. Less commonly, power steering fluid, brake fluid, and other hydraulic fluids may be cooled by an auxiliary radiator on a vehicle.

Turbo charged or supercharged engines may have an intercooler, which is an air-to-air or air-to-water radiator used to cool the incoming air charge—not to cool the engine.

Aircraft

Aircraft with liquid-cooled piston engines (usually inline engines rather than radial) also require radiators. As airspeed is higher than for cars, these are efficiently cooled in flight, and so do not require large areas or cooling fans. Many high-performance aircraft however suffer extreme overheating problems when idling on the ground - a mere 7 minutes for a Spitfire. This is similar to Formula 1 cars of today, when stopped on the grid with engines running they require ducted air forced into their radiator pods to prevent overheating.

Surface Radiators

Reducing drag is a major goal in aircraft design, including the design of cooling systems. An early technique was to take advantage of an aircraft 's abundant airflow to replace the honeycomb core (many surfaces, with a high ratio of surface to volume) by a surface mounted radiator. This uses a single surface blended into the fuselage or wing skin, with the coolant flowing through pipes at the back of this surface. Such designs were seen mostly on World War I aircraft.

As they are so dependent on airspeed, surface radiators are even more prone to overheating when ground-running. Racing aircraft such as the Supermarine S.6B, a racing seaplane with radiators built into the upper surfaces of its floats, have been described as "being flown on the temperature gauge" as the main limit on their performance.

Surface radiators have also been used by a few high-speed racing cars, such as Malcolm Campbell's Blue Bird of 1928.

Pressurized Cooling Systems

It is generally a limitation of most cooling systems that the cooling fluid not be allowed to boil, as the need to handle gas in the flow greatly complicates design. For a water cooled system, this means that the maximum amount of heat transfer is limited by the specific heat capacity of water and the difference in temperature between ambient and 100 °C. This provides more effective cooling in the winter, or at higher altitudes where the temperatures are low.

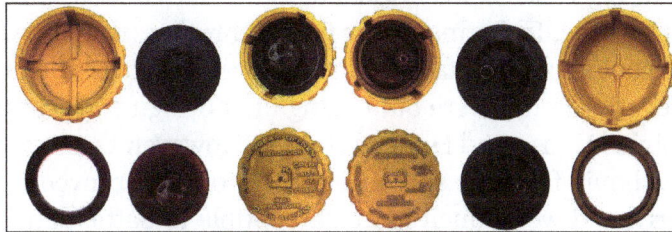

Radiator caps for pressurized automotive cooling systems. Of the two valves, one prevents the creation of a vacuum, the other limits the pressure.

Another effect that is especially important in aircraft cooling is that the specific heat capacity changes with pressure and this pressure changes more rapidly with altitude than the drop in temperature. Thus, generally, liquid cooling systems lose capacity as the aircraft climbs. This was a major limit on performance during the 1930s when the introduction of turbosuperchargers first allowed convenient travel at altitudes above 15,000 ft, and cooling design became a major area of research.

The most obvious, and common, solution to this problem was to run the entire cooling system under pressure. This maintained the specific heat capacity at a constant value, while the outside air temperature continued to drop. Such systems thus improved cooling capability as they climbed. For most uses, this solved the problem of cooling high-performance piston engines, and almost all liquid-cooled aircraft engines of the World War II period used this solution.

However, pressurized systems were also more complex, and far more susceptible to damage - as the cooling fluid was under pressure, even minor damage in the cooling system like a single rifle-calibre bullet hole, would cause the liquid to rapidly spray out of the hole. Failures of the cooling systems were, by far, the leading cause of engine failures.

Evaporative Cooling

Although it is more difficult to build an aircraft radiator that is able to handle steam, it is by no means impossible. The key requirement is to provide a system that condenses the steam back into liquid before passing it back into the pumps and completing the cooling loop. Such a system can take advantage of the specific heat of vaporization, which in the case of water is five times the specific heat capacity in the liquid form. Additional gains may be had by allowing the steam to become superheated. Such systems, known as evaporative coolers, were the topic of considerable research in the 1930s.

Consider two cooling systems that are otherwise similar, operating at an ambient air temperature of 20 °C. An all-liquid design might operate between 30 °C and 90 °C, offering 60 °C of temperature difference to carry away heat. An evaporative cooling system might operate between 80 °C and 110 °C, which at first glance appears to be much less temperature difference, but this analysis overlooks the enormous amount of heat energy soaked up during the generation of steam, equivalent to 500 °C. In effect, the evaporative version is operating between 80 °C and 560 °C, a 480 °C effective temperature difference. Such a system can be effective even with much smaller amounts of water.

The downside to the evaporative cooling system is the *area* of the condensers required to cool the steam back below the boiling point. As steam is much less dense than water, a correspondingly larger surface area is needed to provide enough airflow to cool the steam back down. The Rolls-Royce Goshawk design of 1933 used conventional radiator-like condensers and this design proved to be a serious problem for drag. In Germany, the Günter brothers developed an alternative design combining evaporative cooling and surface radiators spread all over the aircraft wings, fuselage and even the rudder. Several aircraft were built using their design and set numerous performance records, notably the Heinkel He 119 and Heinkel He 100. However, these systems required numerous pumps to return the liquid from the spread-out radiators and proved to be extremely difficult to keep running properly, and were much more susceptible to battle damage. Efforts to develop this system had generally been abandoned by 1940. The need for evaporative cooling was soon to be negated by the widespread availability of ethylene glycol based coolants, which had a lower specific heat, but a much higher boiling point than water.

Radiator Thrust

An aircraft radiator contained in a duct heats the air passing through, causing the air to expand and gain velocity. This is called the Meredith effect, and high-performance piston aircraft with well-designed low-drag radiators (notably the P-51 Mustang) derive thrust from it. The thrust was significant enough to offset the drag of the duct the radiator was enclosed in and allowed the aircraft to achieve zero cooling drag. At one point, there were even plans to equip the Spitfire with an afterburner, by injecting fuel into the exhaust duct after the radiator and igniting it. Afterburning is achieved by injecting additional fuel into the engine downstream of the main combustion cycle.

Stationary Plant

Engines for stationary plant are normally cooled by radiators in the same way as automobile engines. However, in some cases, evaporative cooling is used via a cooling tower.

Wax Thermostatic Element

Car engine wax thermostatic element.

The wax thermostatic element was invented in 1934 by Sergius Vernet. Its principal application is in automotive thermostats used in the engine cooling system. The first applications in the plumbing and heating industries were in Sweden and in Switzerland.

Wax thermostatic elements transform heat energy into mechanical energy using the thermal expansion of waxes when they melt. This wax motor principle also finds applications besides engine cooling systems, including heating system thermostatic radiator valves, plumbing, industrial, and agriculture.

Automotive Thermostats

The internal combustion engine cooling thermostat maintains the temperature of the engine near its optimum operating temperature by regulating the flow of coolant to an air cooled radiator. This regulation is now carried out by an internal thermostat. Conveniently, both the sensing element of the thermostat and its control valve may be placed at the same location, allowing the use of a simple self-contained non-powered thermostat as the primary device for the precise control of engine temperature. Although most vehicles now have a temperature-controlled electric cooling fan, "the unassisted air stream can provide sufficient cooling up to 95% of the time" and so such a fan is not the mechanism for primary control of the internal temperature.

Research in the 1920s showed that cylinder wear was aggravated by condensation of fuel when it contacted a cool cylinder wall which removed the oil film. The development of the automatic thermostat in the 1930s solved this problem by ensuring fast engine warm-up.

The first thermostats used a sealed capsule of an organic liquid with a boiling point just below the desired opening temperature. These capsules were made in the form of a cylindrical bellows. As the liquid boiled inside the capsule, the capsule bellows expanded, opening a sheet brass plug valve within the thermostat. As these thermostats could fail in service, they were designed for easy replacement during servicing, usually by being mounted under the water outlet fitting at the top of the cylinder block. Conveniently this was also the hottest accessible part of the cooling circuit, giving a fast response when warming up.

Cooling circuits have a small bypass path even when the thermostat is closed, usually by a small hole in the thermostat. This allows enough flow of cooling water to heat the thermostat when warming up. It also provided an escape route for trapped air when first filling the system. A larger bypass is often provided, through the cylinder block and water pump, so as to keep the rising temperature distribution even.

Work on cooling high-performance aircraft engines in the 1930s led to the adoption of pressurised cooling systems, which became common on post-war cars. As the boiling point of water increases with increasing pressure, these pressurised systems could run at a higher temperature without boiling. This increased both the working temperature of the engine, thus its efficiency, and also the heat capacity of the coolant by volume, allowing smaller cooling systems that required less pump power. A drawback to the bellows thermostat was that it was also sensitive to pressure changes, thus could sometimes be forced shut again by pressure, leading to overheating. The later wax pellet type has a negligible change in its external volume, thus is insensitive to pressure changes. It is otherwise identical in operation to the earlier type. Many cars of the 1950s, or earlier, that were originally built with bellows thermostats were later serviced with replacement wax capsule thermostats, without requiring any change or adaption.

This most common modern form of thermostat now uses a wax pellet inside a sealed chamber. Rather than a liquid-vapour transition, these use a solid-liquid transition, which for waxes is accompanied by a large increase in volume. The wax is solid at low temperatures, and as the engine heats up, the wax melts and expands. The sealed chamber operates a rod which opens a valve when the operating temperature is exceeded. The operating temperature is fixed, but is determined by the specific composition of the wax, so thermostats of this type are available to maintain different temperatures, typically in the range of 70 to 90 °C (160 to 200 °F). Modern engines run hot, that is, over 80 °C (180 °F), in order to run more efficiently and to reduce the emission of pollutants.

While the thermostat is closed, there is no flow of coolant in the radiator loop, and coolant water is instead redirected through the engine, allowing it to warm up rapidly while also avoiding hot spots. The thermostat stays closed until the coolant temperature reaches the nominal thermostat opening temperature. The thermostat then progressively opens as the coolant temperature increases to the optimum operating temperature, increasing the coolant flow to the radiator. Once the optimum operating temperature is reached, the thermostat progressively increases or decreases its opening in response to temperature changes, dynamically balancing the coolant recirculation flow and coolant flow to the radiator to maintain the engine temperature in the optimum range as engine heat output, vehicle speed, and outside ambient temperature change. Under normal operating conditions the thermostat is open to about half of its stroke travel, so that it can open further or reduce its opening to react to changes in operating conditions. A correctly designed thermostat will never be fully open or fully closed while the engine is operating normally, or overheating or overcooling would occur.

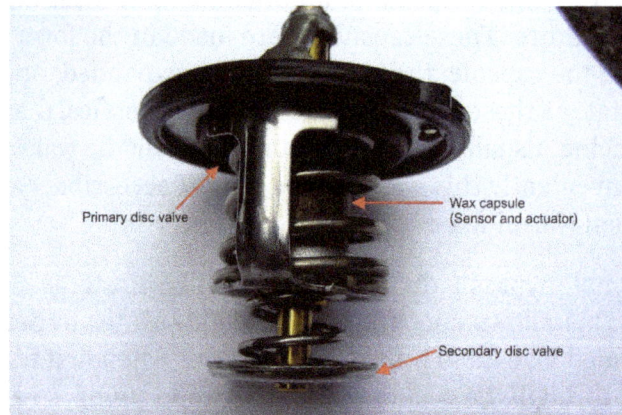

Double valve engine thermostat.

Engines which require a tighter control of temperature, as they are sensitive to "Thermal shock" caused by surges of coolant, may use a "constant inlet temperature" system. In this arrangement the inlet cooling to the engine is controlled by double-valve thermostat which mixes a re-circulating sensing flow with the radiator cooling flow. These employ a single capsule, but have two valve discs. Thus a very compact, and simple but effective, control function is achieved.

The wax used within the thermostat is specially manufactured for the purpose. Unlike a standard paraffin wax, which has a relatively wide range of carbon chain lengths, a wax used in the thermostat application has a very narrow range of carbon molecule chains. The extent of the chains is usually determined by the melting characteristics demanded by the specific end application. To manufacture a product in this manner requires very precise levels of distillation.

Types of Elements

Flat Diaphragm Element

The temperature sensing material contained in the cup transfers pressure to the piston by means of the diaphragm and the plug, held tightly in position by the guide. On cooling, the initial position of the piston is obtained by means of a return spring. Flat diaphragm elements are particularly noted for their high level of accuracy, and therefore mainly used in sanitary installations and heating.

Squeeze-push Elements

Squeeze-Push elements contain a synthetic rubber sleeve-like component shaped like the 'finger of a glove' which surrounds the piston. As the temperature increases, pressure from the expansion of the thermostatic material moves the piston with a lateral squeeze and a vertical push. As with the flat diaphragm element, the piston returns to its initial position by means of a return spring. These elements are slightly less accurate but provide a longer stroke.

Properties

The stroke is the movement of the piston in relation to its starting point. The ideal stroke corresponds to the temperature range of the elements. According to the type of element, it can vary from 1.5 mm to 16 mm.

The temperature range lies between the minimum and maximum operating temperature of the element. Elements can cover temperatures ranging from -15 °C to +120 °C. Elements may move in proportion to the temperature change over some part of the range, or may open suddenly around a particular temperature depending on the composition of the waxes.

Hysteresis is the difference noted between the upstroke and down stroke curve on heating and cooling of the element. Hysteresis is caused by the thermal inertia of the element and by the friction between the parts in motion

Pressure Cap and Reserve Tank

As coolant gets hot, it expands. Since the cooling system is sealed, this expansion causes an increase

in pressure in the cooling system, which is normal and part of the design. When coolant is under pressure, the temperature where the liquid begins to boil is considerably higher. This pressure, coupled with the higher boiling point of ethylene glycol, allows the coolant to safely reach temperatures in excess of 250 degrees.

The radiator pressure cap is a simple device that will maintain pressure in the cooling system up to a certain point. If the pressure builds up higher than the set pressure point, there is a spring loaded valve, calibrated to the correct Pounds per Square Inch (psi), to release the pressure.

When the cooling system pressure reaches the point where the cap needs to release this excess pressure, a small amount of coolant is bled off. It could happen during stop and go traffic on an extremely hot day, or if the cooling system is malfunctioning. If it does release pressure under these conditions, there is a system in place to capture the released coolant and store it in a plastic tank that is usually not pressurized. Since there is now less coolant in the system, as the engine cools down a partial vacuum is formed. The radiator cap on these closed systems has a secondary valve to allow the vacuum in the cooling system to draw the coolant back into the radiator from the reserve tank (like pulling the plunger back on a hypodermic needle). There are usually markings on the side of the plastic tank marked Full-Cold, and Full Hot. When the engine is at normal operating temperature, the coolant in the translucent reserve tank should be up to the Full-Hot line. After the engine has been sitting for several hours and is cold to the touch, the coolant should be at the Full-Cold line.

Water Pump

A water pump is a simple device that will keep the coolant moving as long as the engine is running. It is usually mounted on the front of the engine and turns whenever the engine is running. The water pump is driven by the engine through one of the following:

- A fan belt that will also be responsible for driving an additional component like an alternator or power steering pump.

- A serpentine belt, which also drives the alternator, power steering pump and AC compressor among other things.

- The timing belt that is also responsible for driving one or more camshafts.

The water pump is made up of a housing, usually made of cast iron or cast aluminum and an impeller mounted on a spinning shaft with a pulley attached to the shaft on the outside of the pump body. A seal keeps fluid from leaking out of the pump housing past the spinning shaft. The impeller uses centrifugal force to draw the coolant in from the lower radiator hose and send it under pressure into the engine block. There is a gasket to seal the water pump to the engine block and prevent the flowing coolant from leaking out where the pump is attached to the block.

Bypass System

This is a passage that allows the coolant to bypass the radiator and return directly back to the engine. Some engines use a rubber hose, or a fixed steel tube. In other engines, there is a cast in passage built into the water pump or front housing. In any case, when the thermostat is closed, coolant is directed to this bypass and channeled back to the water pump, which sends the coolant back into the engine without being cooled by the radiator.

Freeze Plugs

When an engine block is manufactured, a special sand is molded to the shape of the coolant passages in the engine block. This sand sculpture is positioned inside a mold and molten iron or aluminum is poured to form the engine block. When the casting is cooled, the sand is loosened and removed through holes in the engine block casting leaving the passages that the coolant flows through. Obviously, if we don't plug up these holes, the coolant will pour right out.

Plugging these holes is the job of the freeze-out plug. These plugs are steel discs or cups that are press fit in the holes in the side of the engine block and normally last the life of the engine with no problems. But there is a reason they are called freeze-out plugs. In the early days, many people used plain water in their engines, usually after replacing a burst hose or other cooling system repair.

Needless to say, people are forgetful and many a motor suffered the fate of the water freezing inside the block. Often, when this happened the pressure of the water freezing and expanding forced the freeze-out plugs to pop out, relieving the pressure and saving the engine block from cracking.

(although, just as often the engine cracked anyway). Another reason for these plugs to fail was the fact that they were made of steel and would easily rust through if the vehicle owner was careless about maintaining the cooling system. Antifreeze has rust inhibitors in the formula to prevent this from happening, but those chemicals would lose their effect after 3 years, which is why antifreeze needs to be changed periodically. The fact that some people left plain water in their engines greatly accelerated the rusting of these freeze plugs.

When a freeze plug becomes so rusty that it perforates, you have a coolant leak that must be repaired by replacing the rusted out freeze plug with a new one. This job ranges from fairly easy to extremely difficult depending on the location of the affected freeze plug. Freeze plugs are located on the sides of the engine, usually 3 or 4 per side. There are also freeze plugs on the back of the engine on some models and also on the heads.

As long as you are good about maintaining the cooling system, you need never worry about these plugs failing on modern vehicles.

Head Gaskets and Intake Manifold Gaskets

All internal combustion engines have an engine block and one or two cylinder heads. The mating surfaces where the block and head meet are machined flat for a close, precision fit, but no amount of careful machining will allow them to be completely water tight or be able to hold back combustion gases from escaping past the mating surfaces.

In order to seal the block to the heads, we use a head gasket. The head gasket has several things it needs to seal against. The main thing is the combustion pressure on each cylinder. Oil and coolant must easily flow between block and head and it is the job of the head gasket to keep these fluids from leaking out or into the combustion chamber, or each other for that matter.

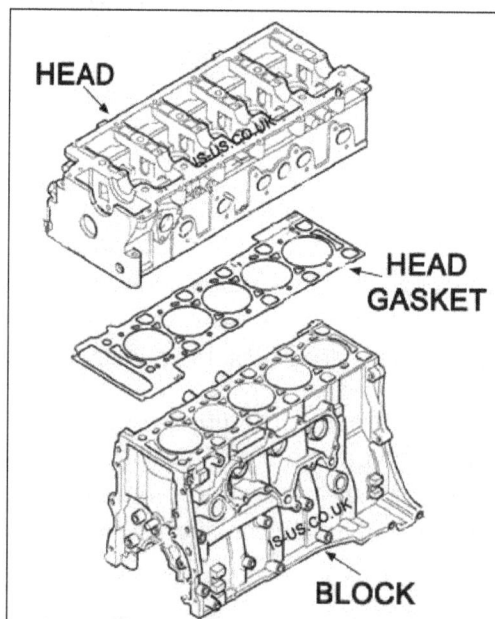

A typical head gasket is usually made of soft sheet metal that is stamped with ridges that surround all leak points. When the head is placed on the block, the head gasket is sandwiched between them.

Many bolts, called head bolts are screwed in and tightened down causing the head gasket to crush and form a tight seal between the block and head.

Head gaskets usually fail if the engine overheats for a sustained period of time causing the cylinder head to warp and release pressure on the head gasket. This is most common on engines with cast aluminum heads, which are now on just about all modern engines.

Once coolant or combustion gases leak past the head gasket, the gasket material is usually damaged to a point where it will no longer hold the seal. This causes leaks in several possible areas. For example:

- Combustion gases could leak into the coolant passages causing excessive pressure in the cooling system.

- Coolant could leak into the combustion chamber causing coolant to escape through the exhaust system, often causing a white cloud of smoke at the tailpipe.

- Other problems such as oil mixing with the coolant or being burned out the exhaust are also possible.

Some engines are more susceptible to head gasket failure than others. I have seen blown head gaskets on engines that just started to overheat and were running hot for less than 5 minutes. The best advice I can give is, if the engine shows signs of overheating, find a place to pull over and shut the engine off as quickly as possible.

Head gaskets themselves are relatively cheap, but it is the labor that's the killer. A typical head gasket replacement is a several hour job where the top part of the engine must be completely disassembled. These jobs can easily reach $1,000 or more.

On V type engines, there are two heads, meaning two head gaskets. While the labor won't double if both head gaskets need to be replaced, it will probably add a good 30% more labor to replace both. If only one head gasket has failed, it is usually not necessary to replace both, but it could be added insurance to get them both done at once.

A head gasket replacement begins with the diagnosis that the head gasket has failed. There is no way for a technician to know for certain whether there is additional damage to the cylinder head or other components without first disassembling the engine. All he or she knows is that fluid and/or combustion is not being contained.

One way to tell if a head gasket has failed is through a combustion leak test on the radiator. This is a chemical test that determines if there are combustion gases in the engine coolant. Another way is to remove the spark plugs and crank the engine while watching for water spray from one or more spark plug holes. Once the technician has determined that a head gasket must be replaced, an estimate is given for parts and labor. The technician will then explain that there may be additional charges after the engine is opened if more damage is found.

Heater Core

The hot coolant is also used to provide heat to the interior of the vehicle when needed. This is a simple and straight forward system that includes a heater core, which looks like a small version of a radiator, connected to the cooling system with a pair of rubber hoses. One hose brings hot coolant from the water pump to the heater core and the other hose returns the coolant to the top of the engine. There is usually a heater control valve in one of the hoses to block the flow of coolant into the heater core when maximum air conditioning is called for.

A fan, called a blower, draws air through the heater core and directs it through the heater ducts to the interior of the car. Temperature of the heat is regulated by a blend door that mixes cool outside air, or sometimes air conditioned air with the heated air coming through the heater core. This blend door allows you to control the temperature of the air coming into the interior. Other doors allow you to direct the warm air through the ducts on the floor, the defroster ducts at the base of the windshield, and the air conditioning ducts located in the instrument panel.

Hoses

There are several rubber hoses that make up the plumbing to connect the components of the cooling system. The main hoses are called the upper and lower radiator hoses. These two hoses are approximately 2 inches in diameter and direct coolant between the engine and the radiator. Two additional hoses, called heater hoses, supply hot coolant from the engine to the heater core. These hoses are approximately 1 inch in diameter. One of these hoses may have a heater control valve mounted in-line to block the hot coolant from entering the heater core when the air conditioner is set to max-cool. A fifth hose, called the bypass hose, is used to circulate the coolant through the engine, bypassing the radiator, when the thermostat is closed. Some engines do not use a rubber hose. Instead, they might use a metal tube or have a built-in passage in the front housing.

These hoses are designed to withstand the pressure inside the cooling system. Because of this, they are subject to wear and tear and eventually may require replacing as part of routine maintenance. If the rubber is beginning to look dry and cracked, or becomes soft and spongy, or you notice some ballooning at the ends, it is time to replace them. The main radiator hoses are usually molded to a shape that is designed to rout the hose around obstacles without kinking. When purchasing replacements, make sure that they are designed to fit the vehicle.

There is a small rubber hose that runs from the radiator neck to the reserve bottle. This allows coolant that is released by the pressure cap to be sent to the reserve tank. This rubber hose is about a quarter inch in diameter and is normally not part of the pressurized system. Once the engine is cool, the coolant is drawn back to the radiator by the same hose.

References

- Murr, andrew (9 january 2008). "an exhausting new crime — what thieves are stealing from today's cars". Newsweek. Retrieved 7 january 2011

- Castaignède, laurent (2018). Airvore ou la face obscure des transports ; chronique d'une pollution annoncée. Montréal (québec): écosociété. Pp. 109–110 and illustration p. 7. Isbn 9782897193591. Oclc 1030881466

- How-does-a-muffler-work: yourmechanic.com, Retrieved 21 May, 2019

- Palucka, tim (winter 2004). "doing the impossible". Invention & technology. 19 (3). Archived from the original on 3 december 2008. Retrieved 14 december 2011

- Head-gaskets-and-intake-manifold-gaskets, coolingsystem, classroom: carparts.com, Retrieved 8 January, 2019

Permissions

All chapters in this book are published with permission under the Creative Commons Attribution Share Alike License or equivalent. Every chapter published in this book has been scrutinized by our experts. Their significance has been extensively debated. The topics covered herein carry significant information for a comprehensive understanding. They may even be implemented as practical applications or may be referred to as a beginning point for further studies.

We would like to thank the editorial team for lending their expertise to make the book truly unique. They have played a crucial role in the development of this book. Without their invaluable contributions this book wouldn't have been possible. They have made vital efforts to compile up to date information on the varied aspects of this subject to make this book a valuable addition to the collection of many professionals and students.

This book was conceptualized with the vision of imparting up-to-date and integrated information in this field. To ensure the same, a matchless editorial board was set up. Every individual on the board went through rigorous rounds of assessment to prove their worth. After which they invested a large part of their time researching and compiling the most relevant data for our readers.

The editorial board has been involved in producing this book since its inception. They have spent rigorous hours researching and exploring the diverse topics which have resulted in the successful publishing of this book. They have passed on their knowledge of decades through this book. To expedite this challenging task, the publisher supported the team at every step. A small team of assistant editors was also appointed to further simplify the editing procedure and attain best results for the readers.

Apart from the editorial board, the designing team has also invested a significant amount of their time in understanding the subject and creating the most relevant covers. They scrutinized every image to scout for the most suitable representation of the subject and create an appropriate cover for the book.

The publishing team has been an ardent support to the editorial, designing and production team. Their endless efforts to recruit the best for this project, has resulted in the accomplishment of this book. They are a veteran in the field of academics and their pool of knowledge is as vast as their experience in printing. Their expertise and guidance has proved useful at every step. Their uncompromising quality standards have made this book an exceptional effort. Their encouragement from time to time has been an inspiration for everyone.

The publisher and the editorial board hope that this book will prove to be a valuable piece of knowledge for students, practitioners and scholars across the globe.

Index

A

Adiabatic Efficiency, 129
Aerodynamic Drag, 8, 124-125, 141
Atkinson Cycle, 28, 33-34, 36-37
Aviation Gasoline, 127

B

Brayton Cycle, 66, 78-79
Bypass Stream, 69, 72-73, 76-77

C

Capacity Rating, 118
Carbon Monoxide, 5, 30, 46, 55, 144, 190, 192-194
Carburetor, 2, 19, 38, 57, 64-65, 100, 125, 144, 158-163, 166-168, 183
Carnot Cycle, 44
Cast Iron, 105-106, 113, 219
Centrifugal Compressor, 82, 87, 130, 133
Chemical Energy, 1, 7, 9-10, 44
Combustion Chamber, 1-2, 14-15, 17-19, 21-22, 29-31, 37, 42, 46, 58, 67, 75, 77, 79, 82, 102, 105, 108-109, 115, 128, 136, 138, 159, 164, 174, 196-197, 201, 206, 220-221
Compression Ratios, 4, 23, 38-39, 44, 96, 165
Compression Stroke, 2-4, 10-11, 22, 27, 33-35, 37-38, 44, 57, 65-66, 109, 143, 173-174, 181-183
Connecting Rod, 26, 51, 65, 106-107, 110-113
Continuous Combustion, 78
Crankshaft, 1, 3, 8-11, 13, 16, 19-20, 22, 25-27, 32-34, 38-40, 42, 44, 54, 56-57, 63, 65, 87, 97-99, 101-102, 106-107, 110-112, 116, 123, 128, 143, 153, 161, 179, 182, 188, 209
Creep Resistance, 81
Cross Scavenging, 21, 114-115
Crossflow Scavenging, 53
Crosshead Pistons, 113

D

Diesel Locomotives, 16, 61
Diesel Particulate Matter, 55

E

Electronic Control Unit, 47, 139, 146, 169
Engine Knocking, 131
Ethylene Glycol, 39, 200, 207, 214, 218

Exhaust Gas Recirculation, 5, 27, 56, 146-147, 194
Exhaust Valve, 2, 10, 16, 22, 31, 35, 44 65-66, 100, 106
Expansion Chamber, 13, 15, 17

F

Flat Engines, 39, 199
Foil Bearings, 80, 96, 136
Forced Induction, 18, 37, 128, 131, 141, 146, 148
Four-stroke Cycle, 3, 9-10, 24, 32-33, 50, 181

G

Gas Turbines, 1, 6, 40, 78, 80-86, 89-90, 92-97
Gear Train, 47, 143
General Aviation, 7, 62, 127, 142
Gudgeon Pin, 110, 112-114

I

Indirect-injection Diesel, 4
Inline Engines, 98, 212
Internal Combustion Engine, 1, 4, 7, 11, 25, 27-28, 33, 38, 41, 43, 64, 78, 97, 102, 106, 108-109, 112-113, 116, 119, 128, 141, 146, 154, 156, 158, 171, 188, 190, 199, 206, 215
Isentropic Process, 119

J

Jet Engine, 7, 66, 68-70, 74, 92

K

Kinetic Energy, 27, 45, 87, 130, 133

M

Mass Flow, 68, 72, 105, 123, 146-147, 156
Mechanical Energy, 7, 9-10, 37, 66, 79, 215

N

Naval Vessels, 92-93

O

Octane Rating, 23, 43, 127-128
Otto Cycle Engine, 33, 98
Overrunning Clutch, 121, 143

P

Piston Pin, 106-107, 110
Piston Ring, 24, 116

Poppet Valve, 18
Power-to-weight Ratio, 8-9, 11, 30, 92, 96, 126, 143

R
Radial Configuration, 97, 99
Radial Engine, 8, 98, 127
Ram Pressure, 74-75, 77
Reciprocating Engines, 7-8, 96-97, 111
Reed Valve, 13-14
Rotary Motion, 8, 65, 106
Rotary Valve, 14, 20, 164
Rotatory Engine, 7, 97

S
Schnuerle Porting, 21-22, 115
Spark-ignition Engine, 4-5, 7, 44, 64, 158, 173
Stationary Diesel Engines, 63
Steam Engines, 34, 42, 61, 116, 205
Supersonic Flight, 73-77

T
Thermal Barrier Coatings, 64, 96, 206
Thermal Efficiency, 3, 30, 33, 39-42, 83-85, 96, 123, 210

Thermodynamic Efficiency, 12, 28, 31
Throttle Response, 123, 131, 133, 160
Thrust Bearings, 20, 80
Top Dead Center, 14-15, 17, 20, 22, 38, 166, 168, 170
Torque Controlling, 47, 56
Turbine Engine, 45, 80-81, 87-88, 90-91, 94, 96
Turbo Lag, 26, 90-91, 123, 131-132, 134-135, 137-138
Turbochargers, 4, 18, 57, 85, 91, 121-123, 127-129, 132-135, 137, 140, 142-143
Turbofan Engines, 7, 72-73
Turboprop Engine, 71, 80-81

U
Utilite Engine, 33, 35-36

V
Variable Valve Timing, 33, 100

W
Wankel Engine, 102-103

www.ingramcontent.com/pod-product-compliance
Lightning Source LLC
Chambersburg PA
CBHW082100190326
41458CB00010B/3531